RELATIONSHIP MARKETING

SUCCESSFUL STRATEGIES FOR THE AGE OF THE CUSTOMER

Regis McKenna

Addison-Wesley Publishing Company, Inc.

Reading, Massachusetts • Menlo Park, California • New York
Don Mills, Ontario • Wokingham, England • Amsterdam
Bonn • Sydney • Singapore • Tokyo • Madrid • San Juan
Paris • Seoul • Milan • Mexico City • Taipei

Many of the designations used by manufacturers and sellers to distinguish their products are claimed as trademarks. Where those designations appear in this book and Addison-Wesley was aware of a trademark claim, the designations have been printed in initial capital letters (e.g., Tide).

Library of Congress Cataloging-in-Publication Data

McKenna, Regis.
   Relationship marketing : successful strategies for the age of the customer / Regis McKenna.
     p.   cm.
   Includes index.
   ISBN 0-201-56769-5
   1. Marketing. 2. Customer service. 3. High technology—Marketing. I. Title.
HF5415.M2616   1991
658.8'12—dc20   91-20127
                                                                           CIP

Jacket design by Mike Stromberg
Jacket photo by Jeanene Scott
Text design by Wilson Graphics & Design (Kenneth J. Wilson)
Set in 10-point Palatino by CopyRight, Inc.

1 2 3 4 5 6 7 8 9-MA-9594939291
First Printing, November 1991

To Dianne

Contents

The great obstacle to progress is not ignorance but the illusion of knowledge.

Daniel Boorstin

Preface

The Regis Touch was compiled and written during the late 1970s and early 1980s. This was a time some consider the golden age of venture capital and the entrepreneurial-lead technology revolution. During that time, the company that positioned its particular product or technology in the marketplace early became the big winner. Gaining recognition was an objective in itself. Managements focused inward on the technology and on their organizations. Getting public attention constituted marketing. In the personal computer industry, the distribution channel often was considered the only customer. Establishing new technology and new markets required company leaders to become evangelists. Designing technology to meet customer needs often became secondary. For a while, customers were fascinated by the proliferation and variety of technology. Customers were able to do things they hadn't been able to do before at any price. But the "whole product," or "total solution," was seldom delivered. Service was considered synonymous with repair.

Image was considered something that could be fabricated, designed on Madison Avenue, and overlaid on any company. Perception out-distanced reality by a country mile. In the late '80s reality set in. The customer became the center of attention. High-growth companies always have difficulty keeping reality and perception in balance. When your company is successful, the marketplace seems to set expectations for you. It acts as a temptress, luring you to do and say things you feel are required of you. Add to that the uncertainty that prevails over rapid success and equally rapid failure in the high-tech businesses and the obsession most managements have with creating and holding market positions becomes understandable. There seems to be a constant, incessant push to proselytize the world. Often, instead of concentrating on building a credible position, management spends its time striving to meet the demands of perception.

In the '90s, however, reality will create perceptions. Customers will sort out the real and reject the unreal. They will demand reality.

In most industries, marketing is making a transition from the manipulation of customers' minds to serving customers' needs. Companies that foster customer and market relationships are experimenting, adapting, and changing to the new environment. It is a different and much more difficult environment in which to compete than existed in the '70s and early '80s.

Most of my experience has been with, and therefore most of my examples are taken from, high-technology companies and the marketplace. However, I believe that the lessons learned in this intensely competitive and rapidly changing industry lend themselves well to other industries. I was gratified by the number of people from government, law, consumer and industrial industries, education, and other non-tech or low-tech industries who responded to *The Regis Touch*. Their response supported my belief that the rising power of the customer and relationship marketing are universal phenomena.

Relationship Marketing is the result of many discussions I have had with executives from around the world. Business books are dialogues with their readers. I received hundreds of letters, visited hundreds of new and established companies, and talked to hundreds of executives on the subject of marketing. These experiences have given me an appreciation of the changing nature of marketing and its complexity. I saw how companies wrongly perceived marketing as a separate corporate function, isolated from product development, manufacturing, finance, and sales. Many other changes have occurred in marketing as well. The decline of mass marketing and continued refinement of submarkets, the customization of services and products, intense global competition, and the use of new technology tools in business all have lead to a reevaluation of marketing. The world of business has changed faster than most of us can cope with. But operating in this environment can't be played like a guessing game about where to go and what to do next. Businesses must put in place processes that dance with the environment. A new concept of marketing is needed. *Relationship Marketing* has been written to reflect the many changes in the environment and to suggest a marketing process for success in the decade of the 1990s.

The introductory chapter appeared in the January–February 1991 issue of the *Harvard Business Review*. It establishes new paradigms that I foresee emerging in marketing in the next decade. While the underlying themes of positioning and infrastructure development from *The Regis Touch* remain, they have been substantially changed to reflect changes in the marketplace and the growing power of the customer. Much of the book has been rewritten and many new subjects added. In Chapter 4, the chapter on product positioning, I have added a discussion on "the whole product." In Chapter 5 I have presented an analysis of product successes and failures, the differences between being market driven and marketing driven, the value of a brand name, and the formula for product success. Chapter 7 covers communications from monologue to dialogue. I also discuss the phenomena facing public issues today under the heading, "Matter and antimatter." I decided to leave Chapter 11, the Macintosh Story, nearly intact. This chapter provides a valuable history of the vision and goals of the early Macintosh managers. I felt it would be valuable to see how a product adapts as the market and competition changes. The last section of Chapter 11 discusses what worked and what didn't and where the Macintosh is positioned today.

In the Appendix I have added "Test Your Marketing IQ," and a reading list. The reading list certainly is not comprehensive, but it contains some of my favorite, idea-stimulating books and papers.

Acknowledgments

During nearly thirty years in Silicon Valley, I have worked with more than 300 companies and thousands of executives. I have had the opportunity to meet and work with people from many technology-based companies throughout the world. Silicon Valley is a living laboratory. Here one has the opportunity to observe and learn from many talented people. I have been most fortunate to have lived and worked in Silicon Valley during the most exciting of times.

Three individuals have been especially important to me. The first is Don Valentine, head of marketing at Fairchild Semiconductor and National Semiconductor during the 1960s. He is now a partner with Sequoia Capital, a venture capital firm. The other two are Andy Grove, chairman of Intel, and Steve Jobs, founder of Apple and now NeXT. Both these individuals are legends for their accomplishments in the world of high technology.

I have regular exchanges on the subjects of business and marketing with a number of people. They are all good friends and have contributed significantly to my marketing education. John Sculley, a champion of innovation and change, has exhibited an enormous amount of savvy guiding Apple from less than $1 billion in revenue to almost $6 billion by the end of the 1980s. Michael Spindler, Apple's new president, has been a friend for many years. He is a brilliant business strategist and champion of global marketing. Bill Campbell, formerly executive vice president sales and marketing at Apple and CEO at Apple's software operation Claris, has now taken over the reigns at GO Corporation, a start-up software company. Bill Miller, recently retired president of Stanford Research Institute, is an intellectual powerhouse. Tom Peters, who originally raised the competitive consciousness of America's industries to "the pursuit of excellence," shares his ideas, thoughts, and observations in the form of notes, pamphlets, slips of paper, and notebooks that continually arrive in my mail. Pierre Lamond, general partner of Sequoia

Capital, and I sit on a number of boards together, which gives us an opportunity to discuss management and marketing issues regularly. On our monthly flight to and from Texas, Pierre reads and critiques many of my articles and papers. Tom Moser, partner at KPMG Peat Marwick, is former head of Peat's high-tech practice and has been an enthusiastic champion of technology enterprise helping high-tech firms around the globe. Alan Webber, editorial director of the *Harvard Business Review,* is a friend and has helped me translate many of my ideas into suitable prose for the journal. I also have had the good fortune to know and discuss various marketing issues with Ted Levitt of Harvard Business School. He is one of the very few authors I have found who speaks a universal language on marketing.

Each Monday I have the good fortune to meet with my venture capital partners at Kleiner Perkins Caufield & Byers. These sessions are like being in an advanced classroom for entrepreneurs. The "professors" all have had highly successful careers as entrepreneurs or as operating managers. They include Tom Perkins, Frank Caufield, Brook Byers, John Doerr, Floyd Kvamme, Jim Lally, Vinod Khosla, Bernie Lacroute, and Joe Lacob.

I am indebted to Gayle Holste, my assistant for the past thirteen years. Gayle has provided production help, coordination with my publisher, and valued advice on my writings. My thanks to Mitch Resnick, editor of *The Regis Touch,* and MaryAnn Gutoff, my researcher, writing assistant, and editor. I am grateful to the many people at Regis McKenna, Inc. (RMI), who deserve a great deal of credit for their continued and enthusiastic efforts to solve clients' marketing problems. Thanks also to Ann Dilworth, my publisher at Addison-Wesley.

In 1972 I met Bob Noyce for the first time. He had already invented the integrated circuit, after having founded Fairchild Semiconductor. He was then president of his second start-up, Intel Corporation. It was Bob Noyce, along with seven other associates, who left Shockley Semiconductor to found Fairchild in 1957. The company was backed by venture and corporate financing. Bob and his associates established an egalitarian management style and broadly distributed ownership to their employees. By 1987 some 100 companies in Silicon Valley had traced their origins back to Fairchild. Almost every company in Silicon Valley can trace its origins to the integrated circuit or to the model the Noyce team established for start-up companies. With two revolutionary technology companies under his belt, Bob went on to become the most visible spokesman and champion of American competitiveness. At 58, Bob became the head of Sematech, a consortium of U.S. semiconductor companies aimed at keeping America at the forefront of research. On June 3, 1990, Bob Noyce died. Few people realize how much their lives have been affected by the innovations he was directly or indirectly responsible for. Bob was not only an innovator, he was also a sustainer, and he spent as much time in front of customers as he did in the lab. He saw his ideas through to ensuring successful customer relationships. Everyone who met Bob Noyce—employees, competitors, associates, and customers— felt they had a relationship with him—I did.

A strategy that doesn't speak explicitly about customers and the competitive environment will surely fail to generate and sustain a proper level of customer and competitive consciousness in your company, especially in the important nooks and crannies where the real work gets done.

Theodore Levitt
Thinking About Management

Chapter 1 | **Marketing Is Everything**

The 1990s will belong to the customer. And that is great news for the marketer.

Technology is transforming choice, and choice is transforming the marketplace. As a result, we are witnessing the emergence of a new marketing paradigm—not a "do more" marketing that simply turns up the volume on the sales spiels of the past but a knowledge- and experience-based marketing that represents the once-and-for-all death of the salesman.

Marketing's transformation is driven by the enormous power and ubiquitous spread of technology. So pervasive is technology today that it is virtually meaningless to make distinctions between technology and non-technology businesses and industries: there are *only* technology companies. Technology has moved into products, the workplace, and the marketplace with astonishing speed and thoroughness. Seventy years after they were invented, fractional horsepower motors are in some fifteen to twenty household products in the average American home today. In less than twenty years, the microprocessor has achieved a similar penetration. Twenty years ago, there were fewer than 50,000 computers in use; today more than 50,000 computers are purchased every day.

The defining characteristic of this new technological push is programmability. In a computer chip, programmability means the capability to alter a command, so that one chip can perform a variety of prescribed functions and produce a variety of prescribed outcomes. On the factory floor, programmability transforms the production operation, enabling one machine to produce a wide variety of models and products. More broadly, programmability is the new corporate capability to produce more and more varieties and choices for customers—even to offer each individual customer the chance to design and implement the "program" that will yield the precise product, service, or variety that is right for him or her. The technological promise of programmability has exploded into the reality of almost unlimited choice.

Take the world of drugstores and supermarkets. According to *Gorman's New Product News*, which tracks new product introductions in these two consumer-products arenas, between 1985 and 1989 the number of new products grew by an astonishing 60% to an all-time annual high of 12,055. As venerable a brand as Tide illustrates this multiplication of brand variety. In 1946, Procter & Gamble introduced the laundry detergent, the first ever. For thirty-eight years, one version of Tide served the entire market. Then, in the mid-1980s, Procter & Gamble began to bring out a succession of new Tides: Unscented Tide and Liquid Tide in 1984, Tide with Bleach in 1988, and the concentrated Ultra Tide in 1990.

To some marketers, the creation of almost unlimited customer choice represents a threat—particularly when choice is accompanied by new competitors. Twenty years ago, IBM had only twenty competitors; today it faces more than 5,000, when you count any company that is in the "computer" business. Twenty years ago, there were fewer than ninety semiconductor companies; today there are almost 300 in the United States alone. And not only are the competitors new, bringing with them new products and new strategies, but the customers also are new: 90% of the people who used a computer in 1990 were not using one in 1980. These new customers don't know about the old rules, the old understandings, or the old ways of doing business—and they don't care. What they do care about is a company that is willing to adapt its products or services to fit their strategies. This represents the evolution of marketing to the market-driven company.

Several decades ago, there were sales-driven companies. These organizations focused their energies on changing customers' minds to fit the product—practicing the "any color as long as it's black" school of marketing.

As technology developed and competition increased, some companies shifted their approach and became customer driven. These companies expressed a new willingness to change their product to fit customers' requests—practicing the "tell us what color you want" school of marketing.

In the 1990s, successful companies are becoming market driven, adapting their products to fit their customers' strategies. These companies will practice "let's figure out together whether and how color matters to your larger goal" marketing. It is marketing that is oriented toward creating rather than controlling a market; it is based on developmental education, incremental improvement, and ongoing process rather than on simple market-share tactics, raw sales, and one-time events. Most important, it draws on the base of knowledge and experience that exists in the organization.

These two fundamentals, knowledge-based and experience-based marketing, will increasingly define the capabilities of a successful marketing organization. They will supplant the old approach to marketing and new product development. The old approach—getting an idea, conducting traditional market research, developing a product, testing the market, and finally going to market—is slow, unresponsive, and turf-ridden. Moreover, given the fast-changing marketplace, there is less and less reason to believe that this traditional approach can keep up with real customer wishes and demands or with the rigors of competition.

Consider the much-publicized 1988 lawsuit that Beecham, the international consumer products group, filed against advertising giant Saatchi & Saatchi. The suit, which sought more than $24 million in damages, argued that Yankelovich Clancy Shulman, at that time Saatchi's U.S. market-research subsidiary, had "vastly overstated" the projected market share of a new detergent that Beecham launched. Yankelovich forecast that Beecham's product, Delicare, a cold-water detergent, would win between 45.4% and 52.3% of the U.S. market if Beecham backed it with $18 million of advertising. According to Beecham, however, Delicare's highest market share was 25%; the product generally achieved a market share of between 15% and 20%. The lawsuit was settled out of court, with no clear winner or loser. Regardless of the outcome, however, the issue it illustrates is widespread and fundamental: forecasts, by their very nature, must be unreliable, particularly with technology, competitors, customers, and markets all shifting ground so often, so rapidly, and so radically.

The alternative to this old approach is knowledge-based and experience-based marketing. Knowledge-based marketing requires a company to master a scale of knowledge: of the technology in which it competes; of its competition; of its customers; of new sources of technology that can alter its competitive environment; and of its own organization, capabilities, plans, and way of doing business. Armed with this mastery, companies can put knowledge-based marketing to work in three essential ways: integrating the customer into the design process to guarantee a product that is tailored not only to the customers' needs and desires but also to the customers' strategies; generating niche thinking to use the company's knowledge of channels and markets to identify segments of the market the company can own; and developing the infrastructure of suppliers, vendors, partners, and users whose relationships will help sustain and support the company's reputation and technological edge.

The other half of this new marketing paradigm is experienced-based marketing, which emphasizes interactivity, connectivity, and creativity. With this approach, companies spend time with their customers, constantly monitor their competitors, and develop a feedback-analysis system that turns this information about the market and the competition into important new product intelligence. At the same time, these companies both evaluate their own technology to assess its currency and cooperate with other companies to create mutually advantageous systems and solutions. These close encounters—with customers, competitors, and internal and external technologies—give companies the firsthand experience they need to invest in market development and to take intelligent, calculated risks.

In a time of exploding choice and unpredictable change, marketing—the new marketing—is the answer. With so much choice for customers, companies face the end of loyalty. To combat that threat, they can add sales and marketing people, throwing costly resources at the market as a way to retain customers. But the real solution, of course, is not more marketing but better marketing. And that means marketing that finds a way to integrate the customer into the company, to create and sustain a relationship between the company and the customer.

The marketer must be the integrator, both internally—synthesizing technological capability with market needs—and externally bringing the customer into the company as a participant in the development and adaptation of goods and services. It is a fundamental shift in the role and purpose of marketing: from manipulation of the customer to genuine customer involvement; from telling and selling to communicating and sharing knowledge; from last-in-line function to corporate-credibility champion.

Playing the integrator requires the marketer to command credibility. In a marketplace characterized by rapid change and potentially paralyzing choice, credibility becomes the company's sustaining value. The character of its management, the strength of its financials, the quality of its innovations, the congeniality of its customer references, the capabilities of its alliances—these are the measures of a company's credibility. They are measures that, in turn, directly affect its capacity to attract quality people, generate new ideas, and form quality relationships.

The relationships are the key, the basis of customer choice and company adaptation. After all, what is a successful brand but a special relationship? And who better than a company's marketing people to create, sustain, and interpret the relationship between the company, its suppliers, and its customers? That is why, as the demands on the company have shifted from controlling costs to competing on products to serving customers, the center of gravity in the company has shifted from finance to engineering—and now to marketing. In the 1990s, marketing will do more than sell. It will define the way a company does business.

MARKETING IS EVERYTHING AND EVERYTHING IS MARKETING

The old notion of marketing was epitomized by the ritual phone call from the CEO to the corporate headhunter saying, "Find me a good marketing person to run my marketing operation!" What the CEO wanted, of course, was someone who could take on a discrete set of textbook functions that were generally associated with run-of-the-mill marketing. That person would immediately go to Madison Avenue to hire an advertising agency, change the ad campaign, redesign the company logo, redo the brochures, train the sales force, retain a high-powered public relations firm, and alter or otherwise reposition the company's image.

Behind the CEO's call for "a good marketing person" were a number of assumptions and attitudes about marketing: that it is a distinct function in the company, separate from and usually subordinate to the core functions; that its job is to identify groups of potential customers and find ways to convince them to buy the company's product or service; and that at the heart of it is image making—creating and projecting a false sense of the company and its offerings to lure the customer into the company's grasp. If those assumptions ever were warranted in the past, however, all three are totally unsupportable and obsolete today.

Marketing today is not a function; it is a way of doing business. Marketing is not a new ad campaign or this month's promotion. Marketing has to be all-pervasive, part of everyone's job description, from the receptionists to the board of directors. Its job is neither to fool the customer nor to falsify the company's image. It is to integrate the customer into the design of the product and to design a systematic process for interaction that will create substance in the relationship.

To understand the difference between the old and the new marketing, compare how two high-tech medical instrument companies recently handled similar customer telephone calls requesting the repair and replacement of their equipment. The first company—call it Gluco—delivered the replacement instrument to the customer within twenty-four hours of the request, no questions asked. The box in which it arrived contained instructions for sending back the broken instrument, a mailing label, and even tape to reseal the box. The phone call and the exchange of instruments were handled conveniently, professionally, and with maximum consideration for and minimum disruption to the customer.

The second company—call it Pumpco—handled things quite differently. The person who took the customer's telephone call had never been asked about repairing a piece of equipment; she thoughtlessly sent the customer into the limbo of hold. Finally, she came back on the line to say that the customer would have to pay for the equipment repair and that a temporary replacement would cost an additional $15.

Several days later, the customer received the replacement with no instructions, no information, no directions. Several weeks after the customer returned the broken equipment, it reappeared, repaired but with no instructions concerning the temporary replacement. Finally, the customer got a demand letter from Pumpco, indicating that someone at Pumpco had made the mistake of not sending the equipment C.O.D.

To Pumpco, marketing means selling things and collecting money; to Gluco, marketing means building relationships with its customers. The way the two companies handled two simple customer requests reflects the questions that customers increasingly ask in interactions with all kinds of businesses, from airlines to software makers: Which company is competent, responsive, and well organized? Which company do I trust to get it right? Which company would I rather do business with?

Successful companies realize that marketing is like quality, integral to the organization. Like quality, marketing is an intangible that the customer must experience to appreciate. And like quality—which in the United States has developed from early ideas like planned obsolescence and inspecting quality into more ambitious concepts like the systemization of quality in every aspect of the organization—marketing has been evolutionary.

Marketing has shifted from tricking the customer to blaming the customer to satisfying the customer—and now to integrating the customer systematically. As its next move, marketing must permanently shed its reputation for hucksterism and image making and create an award for marketing much like the Malcolm Baldridge National Quality Award. In fact, companies that continue to see marketing as a bag of tricks will lose out in short order to companies that stress substance and real performance.

Marketing's ultimate assignment is to serve customers' real needs and to communicate the substance of the company—not to introduce the kinds of cosmetics that used to typify the auto industry's annual model changes. And because marketing in the 1990s is an expression of the company's character, it necessarily is a responsibility that belongs to the whole company.

THE GOAL OF MARKETING IS TO OWN THE MARKET, NOT JUST TO SELL THE PRODUCT

U. S. companies typically make two kinds of mistakes. Some get caught up in the excitement and drive of making things, particularly new creations. Others become absorbed in the competition of selling things, particularly to increase their market share in a given product line.

Both approaches could prove fatal to a business. The problem with the first is that it leads to an internal focus. Companies can become so fixated on pursuing their R&D agendas that they forget about the customer, the market, the competition. They end up winning recognition as R&D pioneers but lack the more important capability—sustaining their performance and, sometimes, maintaining their independence. Genentech, for example, clearly emerged as the R&D pioneer in biotechnology, only to be acquired by Roche.

The problem with the second approach is that it leads to a market-share mentality, which inevitably translates into undershooting the market. A market-share mentality leads a company to think of its customers as "share points" and to use gimmicks, spiffs, and promotions to eke out a percentage-point gain. It pushes a company to look for incremental, sometimes even minuscule, growth out of existing products or to spend lavishly to launch a new product in a market where competitors enjoy a fat, dominant position. It turns marketing into an expensive fight over crumbs rather than a smart effort to own the whole pie.

The real goal of marketing is to own the market—not just to make or sell products. Smart marketing means defining what whole pie is yours. It means thinking of your company, your technology, your product in a fresh way, a way that begins by defining what you can lead. Because in marketing, what you lead, you own. Leadership is ownership.

When you own the market, you do different things and you do things differently, as do your suppliers and your customers. When you own the market, you develop your products to serve that market specifically; you define the standards in that market; you bring into your camp third parties who want to develop their own compatible products or offer you new features or add-ons to augment your product; you get the first look at new ideas that others are testing in that market; you attract the most talented people because of your acknowledged leadership position.

Owning a market can become a self-reinforcing spiral. Because you own the market, you become the dominant force in the field; because you dominate the field, you deepen your ownership of the market. Ultimately, you deepen your relationship with your customers as well, as they attribute more and more leadership qualities to a company that exhibits such an integrated performance.

To own the market, a company starts by thinking of a new way to define a market. Take, for instance, the case of Convex Computer. In 1984, Convex was looking to put a new computer on the market. Because of the existing market segmentation, Convex could have seen its only choice as competing for market share in the predefined markets: in supercomputers where Cray dominated or in minicomputers where Digital Equipment Corporation led. Determined to define a market it could own, Convex created the "mini-supercomputer" market by offering a product with a price/performance ratio between Cray's $5 million to $15 million supercomputers and Digital's $300,000 to $750,000 minicomputers. Convex's product, priced between $500,000 and $800,000, offered technological performance less than that of a full supercomputer and more than that of a minicomputer. Within this new market, Convex established itself as the leader.

Intel did the same thing with its microprocessor. The company defined its early products and market more as computers than semiconductors. Intel offered, in essence, a computer on a chip, creating a new category of products that it could own and lead.

Sometimes owning a market means broadening it; other times, narrowing it. Apple has managed to do both in efforts to create and own a market. Apple first broadened the category of small computers to achieve a leadership position. The market definition started out as hobby computers and had many small players. The next step was the home computer—a market that was also crowded and limiting. To own a market, Apple identified the personal computer, which expanded the market concept and made Apple the undeniable market leader.

In a later move, Apple did the opposite, redefining a market by narrowing its definition. Unquestionably, IBM owned the business market; for Apple, a market-share mentality in that arena would have been pointless. Instead, with technology alliances and marketing correctly defined, Apple created—and owned—a whole new market: desktop publishing. Once inside the corporate world with desktop publishing, Apple could deepen and broaden its relationships with the business customer.

Paradoxically, two important outcomes of owning a market are substantial earnings, which can replenish the company's R&D coffers, and a powerful market position, a beachhead from which a company can grow additional market share by expanding both its technological capabilities and its definition of the market. The greatest practitioners of this marketing approach are Japanese companies in industries like autos, commercial electronics, semiconductors, and computers and communications. Their primary goal is ownership of certain target markets. The keiretsu industrial structure allows them to use all of the market's infrastructure to achieve this; relationships in technology, information, politics, and distribution help the company assert its leadership.

The Japanese strategy is consistent. These companies begin by using basic research from the United States to jump-start new product development. From 1950 to 1978, for example, Japanese companies entered into 32,000 licensing arrangements to acquire foreign technology at an estimated cost of $9 billion. But the United States spent at least fifty times that much to do the original R&D. Next, these Japanese companies push out a variety of products to engage the market and to learn and then focus on dominating the market to force foreign competitors to retreat—leaving them to harvest substantial returns. These huge profits are recycled into a new spiral of R&D, innovation, market creation, and market dominance.

That model of competing, which links R&D, technology, innovation, production, and finance—integrated through marketing's drive to own a market—is the approach that all competitors will take to succeed in the 1990s.

MARKETING EVOLVES AS TECHNOLOGY EVOLVES

In a world of mass manufacturing, the counterpart was mass marketing. In a world of flexible manufacturing, the counterpart is flexible marketing. The technology comes first, the ability to market follows. The technology embodies adaptability, programmability, and customizability; now comes marketing that delivers on those qualities.

Today technology has created the promise of "any thing, any way, any time." Customers can have their own version of virtually any product, including one that appeals to mass identification rather than individuality, if they so desire. Think of a product or an industry where customization is not predominant. The telephone? Originally, Bell Telephone's goal was to place a simple, all-black phone in every home. Today there are more than 1,000 permutations and combinations available, with options running the gamut from different colors and portability to answering machines and programmability—as well as services. There is the further promise of optical fiber and the convergence of computers and communications into a unified industry with even greater technological choice.

How about a venerable product like the bicycle, which appeared originally as a sketch in Leonardo da Vinci's notebooks? According to a recent article in the *Washington Post*, the National Bicycle Industrial Company in Kokubu, Japan builds made-to-order bicycles on an assembly line. The bicycles, fitted to each customer's measurements, are delivered within two weeks of the order—and the company offers 11,231,862 variations on its models, at prices only 10 percent higher than ready-made models.

Even newspapers that report on this technology-led move to customization are themselves increasingly customized. Faced with stagnant circulation, the urban daily newspapers have begun to customize their news, advertising, and even editorial and sports pages to appeal to local suburban readers. The *Los Angeles Times*, for example, has seven zoned editions targeting each of the city's surrounding communities.

 What is at work here is the predominant mathematical formula of today's marketing: variety plus service equals customization. For all of its bandying about as a marketing buzzword, customization is a remarkably direct concept—it is the capacity to deal with a customer in a unique way. Technology makes it increasingly possible to do that, but interestingly, marketing's version of the laws of physics makes it increasingly difficult.

According to quantum physics, things act differently at the micro level. Light is the classic example. When subjected to certain kinds of tests, light behaves like a wave, moving in much the way an ocean wave moves. But in other tests, light behaves more like a particle, moving as a single ball. So, scientists ask, is it a wave or a particle? And when is it which?

Markets and customers operate like light and energy. In fact, like light, the customer is more than one thing at the same time. Sometimes consumers behave as part of a group, fitting neatly into social and psychographic classifications. Other times, the consumer breaks loose and is iconoclastic. Customers make and break patterns: the senior citizen market is filled with older people who intensely wish to act youthful, and the upscale market must contend with wealthy people who hide their money behind the most utilitarian purchases.

Markets are subject to laws similar to those of quantum physics. Different markets have different levels of consumer energy, stages in the market's development where a product surges, is absorbed, dissipates, and dies. A fad, after all, is nothing more than a wave that dissipates and then becomes a particle. Take the much-discussed Yuppie market and its association with certain branded consumer products, like BMWs. After a stage of high customer energy and close identification, the wave has broken. Having been saturated and absorbed by the marketplace, the Yuppie association has faded, just as energy does in the physical world. Sensing the change, BMW no longer sells to the Yuppie lifestyle but now focuses on the technological capabilities of its machines. And Yuppies are no longer the wave they once were; as a market, they are more like particles as they look for more individualistic and personal expressions of their consumer energy.

Of course, since particles can also behave like waves again, it is likely that smart marketers will tap some new energy source, such as values, to recoalesce the young, affluent market into a wave. And technology gives marketers the tools they need, such as database marketing, to discern waves and particles and even to design programs that combine enough particles to form a powerful wave.

The lesson for marketers is much the same as that voiced by Buckminster Fuller for scientists: "Don't fight forces; use them." Marketers who follow and use technology, rather than oppose it, will discover that it creates and leads directly to new market forms and opportunities. Take audio cassettes, tapes, and compact discs. For years, record and tape companies jealously guarded their property. Knowing that home hackers pirated tapes and created their own composite cassettes, the music companies steadfastly resisted the forces of technology—until the Personics System realized that technology was making a legitimate market for authorized, high-quality customized composite cassettes and CDs.

Rather than treating the customer as a criminal, Personics saw a market. Today consumers can design personalized music tapes from the Personics System, a revved-up jukebox with a library of over 5,000 songs. For $1.10 per song, consumers tell the machine what to record. In about ten minutes, the system makes a customized tape and prints out a laser-quality label of the selections, complete with the customer's name and a personalized title for the tape. Launched in 1988, the system has already spread to more than 250 stores. Smart marketers have, once again, allowed technology to create the customizing relationship with the customer.

We are witnessing the obsolescence of advertising. In the old model of marketing, it made sense as part of the whole formula: you sell mass-produced goods to a mass market through mass media. Marketing's job was to use advertising to deliver a message to the consumer in a one-way communication: "Buy this!" That message no longer works, and advertising is showing the effects. In 1989, newspaper advertising grew only 4%, compared with 6% in 1988 and 9% in 1987. According to a study by Syracuse University's John Philip Jones, ad spending in the major media has been stalled at 1.5% of GNP since 1984. Ad agency staffing, research, and profitability have been affected.

Three related factors explain the decline of advertising. First, advertising overkill has started to ricochet back on advertising itself. The proliferation of products has yielded a proliferation of messages: U.S. customers are hit with up to 3,000 marketing messages a day. In an effort to bombard the customer with yet one more advertisement, marketers are squeezing as many voices as they can into the space allotted to them. In 1988, for example, 38% of primetime and 47% of weekday daytime television commercials were only fifteen seconds in duration; in 1984, those figures were 6% and 11% respectively. As a result of the shift to fifteen-second commercials, the number of television commercials has skyrocketed; between 1984 and 1988, prime-time commercials increased by 25%, weekday daytime by 24%.

Predictably, however, a greater number of voices translates into a smaller impact. Customers simply are unable to remember which advertisement pitches which product, much less what qualities or attributes might differentiate one product from another. Very simply, it's a jumble out there.

Take the enormously clever and critically acclaimed series of advertisements for Eveready batteries, featuring a tireless marching rabbit. The ad was so successful that a survey conducted by Video Storyboard Tests, Inc. named it one of the top commercials in 1990 for Duracell, Eveready's top competitor. In fact, a full 40% of those who selected the ad as an outstanding commercial attributed it to Duracell. Partly as a consequence of this confusion, reports indicate that Duracell's market share has grown, while Eveready may have shrunk slightly.

Batteries are not the only market in which more advertising succeeds in spreading more confusion. The same thing has happened in markets like athletic footwear and soda pop, where competing companies have signed up so many celebrity sponsors that consumers can no longer keep straight who is pitching what for whom. In 1989, for example, Coke, Diet Coke, Pepsi, and Diet Pepsi used nearly three dozen movie stars, athletes, musicians, and television personalities to tell consumers to buy more cola. But when the smoke and mirrors had cleared, most consumers couldn't remember whether Joe Montana and Don Johnson drank Coke or Pepsi—or both. Or why it really mattered.

 The second development in advertising's decline is an outgrowth of the first: as advertising has proliferated and become more obnoxiously insistent, consumers have gotten fed up. The more advertising seeks to intrude, the more people try to shut it out. Last year, Disney won the applause of commercial-weary customers when the company announced that it would not screen its films in theaters that showed commercials before the feature. A Disney executive was quoted as saying, "Movie theaters should be preserved as environments where consumers can escape from the pervasive onslaught of advertising." Buttressing its position, the company cited survey data obtained from moviegoers, 90% of whom said they did not want commercials shown in movie theaters and 95% of whom said they did want to see previews of coming attractions.

More recently, after a number of failed attempts, the U.S. Congress responded to the growing concerns of parents and educators over the commercial content of children's television. A new law limits the number of minutes of commercials and directs the Federal Communications Commission both to examine "program-length commercials"—cartoon shows linked to commercial product lines—and to make each television station's contribution to children's educational needs a condition for license renewal. This concern over advertising is mirrored in a variety of arenas from public outcry over cigarette marketing plans targeted at blacks and women to calls for more environmentally sensitive packaging and products.

 The underlying reason behind both of these factors is advertising's dirty little secret: it serves no useful purpose. In today's market, advertising simply misses the fundamental point of marketing—adaptability, flexibility, and responsiveness. The new marketing requires a feedback loop; it is this element that is missing from the monologue of advertising but that is built into the dialogue of marketing. The feedback loop, connecting company and customer, is central to the operating definition of a truly market-driven company: a company that adapts in a timely way to the changing needs of the customer.

Apple is one such company. Its Macintosh computer is regarded as a machine that launched a revolution. At its birth in 1984, industry analysts received it with praise and acclaim. But in retrospect, the first Macintosh had many weaknesses: it had limited, nonexpandable memory, virtually no applications software, and a black-and-white screen. For all those deficiencies, however, the Mac had two strengths that more than compensated: it was incredibly easy to use, and it had a user group that was prepared to praise Mac publicly at its launch and to advise Apple privately on how to improve it. In other words, it had a feedback loop. It was this feedback loop that brought about change in the Mac, which ultimately became an open, adaptable, and colorful computer. And it was changing the Mac that saved it.

Months before launching the Mac, Apple gave a sample of the product to 100 influential Americans to use and comment on. It signed up 100 third-party software suppliers who began to envision applications that could take advantage of the Mac's simplicity. It trained over 4,000 dealer salespeople and gave full-day, hands-on demonstrations of the Mac to industry insiders and analysts. Apple got two benefits from this network: educated Mac supporters who could legitimately praise the product to the press and invested consumers who could tell the company what the Mac needed. The dialogue with customers *and* media praise were worth more than any notice advertising could buy.

Apple's approach represents the new marketing model, a shift from monologue to dialogue. It is accomplished through experience-based marketing, where companies create opportunities for customers and potential customers to sample their products and then provide feedback. It is accomplished through beta sites, where a company can install a prelaunch product and study its use and needed refinements. Experienced-based marketing allows a company to work closely with a client to change a product, to adapt the technology—recognizing that no product is perfect when it comes from engineering. This interaction was precisely the approach taken by Xerox in developing its recently announced Docutech System. Seven months before launch, Xerox established twenty-five beta sites. From its prelaunch customers, Xerox learned what adjustments it should make, what service and support it should supply, and what enhancements and related new products it might next introduce.

The goal is adaptive marketing, marketing that stresses sensitivity, flexibility, and resiliency. Sensitivity comes from having a variety of modes and channels through which companies can read the environment, from user groups that offer live feedback to sophisticated consumer scanners that provide data on customer choice in real time. Flexibility comes from creating an organizational structure and operating style that permits the company to take advantage of new opportunities presented by customer feedback. Resiliency comes from learning from mistakes—marketing that listens and responds.

MARKETING A PRODUCT IS MARKETING A SERVICE IS MARKETING A PRODUCT

The line between products and services is fast eroding. What once appeared to be a rigid polarity now has become a hybrid: the servicization of products and the productization of services. When General Motors makes more money from lending its customers money to buy its cars than it makes from manufacturing the cars, is it marketing its products or its services? When IBM announces to all the world that it is now in the systems-integration business—the customer can buy any box from any vendor and IBM will supply the systems know-how to make the whole thing work together—is it marketing its products or its services? In fact, the computer business today is 75% services; it consists overwhelmingly of applications knowledge, systems analysis, systems engineering, systems integration, networking solutions, security, and maintenance.

The point applies just as well to less grandiose companies and to less expensive consumer products. Take the large corner drugstore that stocks thousands of products, from cosmetics to wristwatches. The products are for sale, but the store is actually marketing a service—the convenience of having so much variety collected and arrayed in one location. Or take any of the ordinary products found in the home, from boxes of cereal to table lamps to VCRs. All of them come with some form of information designed to perform a service: nutritional information to indicate the actual food value of the cereal to the health-conscious consumer; a United Laboratories label on the lamp as an assurance of testing; an operating manual to help the nontechnical VCR customer rig up the new unit. There is ample room to improve the quality of this information—to make it more useful, more convenient, or even more entertaining—but in almost every case, the service information is a critical component of the product.

On the other side of the hybrid, service providers are acknowledging the productization of services. Service providers, such as banks, insurance companies, consulting firms, even airlines and radio stations, are creating tangible events, repetitive and predictable exercises, standard and customizable packages that are product services. A frequent-flier or a frequent-listener club is a product service, as are regular audits performed by consulting firms or new loan packages assembled by banks to respond to changing economic conditions.

As products and services merge, it is critical for marketers to understand clearly what marketing the new hybrid is not. The service component is not satisfied by repairing a product if it breaks. Nor is it satisfied by an 800 number, a warranty, or a customer survey form. What customers want most from a product is often qualitative and intangible; it is the service that is integral to the product. Service is not an event; it is the process of creating a customer environment of information, assurance, and comfort. Consider an experience that by now must have become commonplace for all of us as consumers. You go to an electronics store and buy an expensive piece of audio or video equipment, say, a CD player, a VCR, or a video camera. You take it home, and a few days later, you accidentally drop it. It breaks. It won't work. Now, as a customer, you have a decision to make. When you take it back to the store, do you say it was broken when you took it out of the box? Or do you tell the truth?

The answer, honestly, depends on how you think the store will respond. But just as honestly, most customers appreciate a store that encourages them to tell the truth by making good on all customer problems. Service is, ultimately, an environment that encourages honesty. The company that adopts a "we'll make good on it, no questions asked" policy in the face of adversity may win a customer for life.

Marketers who ignore the service component of their products focus on competitive differentiation and tools to penetrate markets. Marketers who appreciate the importance of the product-service hybrid focus on building loyal customer relationships.

TECHNOLOGY MARKETS TECHNOLOGY

Technology and marketing once may have looked like opposites. The cold, impersonal sameness of technology and the high-touch, human uniqueness of marketing seemed eternally at odds. Computers would only make marketing less personal; marketing could never learn to appreciate the look and feel of computers, databases, and the rest of the high-tech paraphernalia.

On the grounds of cost, a truce was eventually arranged. Very simply, marketers discovered that real savings could be gained by using technology to do what previously had required expensive, intensive, and often risky, people-directed field operations. For example, marketers learned that by matching a database with a marketing plan to simulate a new product launch on a computer, they could accomplish in ninety days and for $50,000 what otherwise would take as long as a year and cost at least several hundred thousand dollars.

But having moved beyond the simple automation-for-cost-saving stage, technology and marketing have now not only fused but also begun to feed back to each other. The result is the transformation of both technology and the product and the reshaping of both the customer and the company. Technology permits information to flow in both directions between the customer and the company. It creates the feedback loop that integrates the customer into the company, allows the company to own a market, permits customization, creates a dialogue, and turns a product into a service and a service into a product. The direction in which Genentech has moved in its use of laptop and hand-held computers illustrates the transforming power of technology as it merges with marketing. Originally, the biotechnology company planned to have salespeople use laptops on their sales calls as a way to automate the sales function. Sales reps, working solely out of their homes, would use laptops to get and send electronic mail, file reports on computerized "templates," place orders, and receive company press releases and information updates. In addition, the laptops would enable sales reps to keep databases that would track customers' buying histories and company performance. That was the initial level of expectations—very low.

In fact, the technology-marketing marriage has dramatically altered the customer–company relationship and the job of the sales rep. Sales reps have emerged as marketing consultants. Armed with technical information generated and gathered by Genentech, sales reps can provide a valuable educational service to their customers, who are primarily pharmacists and physicians. For example, analysis of the largest study of children with a disease called short stature is available only through Genentech and its representatives. With this analysis, which is based on clinical studies of 6,000 patients between the ages of one month and thirty years, and with the help of an on-line "growth calculator," doctors can better judge when to use the growth hormone Protropin.

Genentech's system also includes a general educational component. Sales reps can use their laptops to access the latest articles or technical reports from medical conferences to help doctors keep up to date. The laptops also make it possible for doctors to use sales reps as research associates: Genentech has a staff of medical specialists who can answer highly technical questions posed through an on-line question-and-answer template. When sales reps enter a question on the template, the e-mail function immediately routes it to the appropriate specialist. For relatively simple questions, on-line answers come back to the sales rep within a day.

In the l990s, Genentech's laptop system—and the hundreds of similar applications that sprang up in the 1980s to automate sales, marketing, service, and distribution—will seem like a rather obvious and primitive way to mold technology and marketing. The marketer will have available not only existing technologies but also their converging capabilities: personal computers, databases, CD-ROMs, graphic displays, multimedia, color terminals, computer-video technology, networking, a custom processor that can be built into anything anywhere to create intelligence on a countertop or a dashboard, scanners that read text, and networks that instantaneously create and distribute vast reaches of information.

As design and manufacturing technologies advance into "real time" processes, marketing will move to eliminate the gap between production and consumption. The result will be marketing workstations—the marketers' counterpart to CAD/CAM systems for engineers and product designers. The marketing workstation will draw on graphic, video, audio, and numeric information from a network of databases. The marketer will be able to look through windows on the workstation and manipulate data, simulate markets and products, bounce concepts off others in distant cities, write production orders for product designs and packaging concepts, and obtain costs, timetables, and distribution schedules.

Just as computer-comfortable children today think nothing of manipulating figures and playing fantastic games on the same color screens, marketers will use the workstation to play both designer and consumer. The workstation will allow marketers to integrate data on historic sales and cost figures, competitive trends, and consumer patterns. At the same time, marketers will be able to create and test advertisements and promotions, evaluate media options, and analyze viewer and readership data. And finally, marketers will be able to obtain instant feedback on concepts and plans and to move marketing plans rapidly into production.

The marriage of technology and marketing should bring with it a renaissance of marketing R&D—a new capability to explore new ideas, to test them against the reactions of real customers in real time, and to advance to experience-based leaps of faith. It should be the vehicle for bringing the customer inside the company and for putting marketing in the center of the company.

In the 1990s, the critical dimensions of the company—including all of the attributes that together define how the company does business—are ultimately the functions of marketing. That is why marketing is everyone's job, why marketing is everything and everything is marketing.

Never sacrifice economies of time to achieve economies of scale.

Freeman Dyson
Infinite in All Directions

Chapter 2 | New Themes for New Marketing

Change is so rapid and unpredictable today that established patterns of market behavior are no longer tenable. Today's management is faced with a myriad of new and changing business circumstances often far beyond its realm of control. In this new competitive era several things happen:

1. Product and service diversity increase even in narrow market segments.

2. Global competition increases.

3. Markets are so segmented that niche becomes king.

4. Industry distinctions are blurred.

5. Product life cycles are accelerated.

6. Distribution channels are in constant flux. While staying close to the customer is paramount, channels often obscure customer dialogue.

7. Traditional promotional media amplify the noise level and fail to communicate clear messages. Confusion reigns.

8. Organizations downsize and restructure, looking for new ways to do business.

9. The business environment and course of competitive events are unpredictable.

10. Forecasting and research don't provide a clear path of action.

It took Apple just seven years, Sun six years, and Compaq five years to reach $1 billion in revenue. Technology-based companies founded before the 1980s took decades to reach that level of revenue. Today's marketing is more time compressed. The gap between product inception and product completion is very narrow. Managers must be clear on what products to pursue and how best to pursue them. And they must act quickly. Experimentation can be catastrophic to a company that makes a wrong decision while the competition is on its heels. Complicating the situation is the fact that technology-based industries are young. There isn't a lot of seasoned expertise among employees, and management practices from one industry don't translate easily into another.

Managing change can't be done in the abstract. It has to be done with one foot in the market and one foot in the technology. That way, decisions can be made knowledgeably, gauging the opportunities of the market while assessing the capabilities and resources of the company. Everyone in management must be acutely aware of market dynamics. Marketing must become a corporate learning process.

TIME TO MARKET AND TIME TO ACCEPTANCE

Time is an ever-present danger to any business. The window of opportunity stays open only briefly. Competitors can step in and build on your mistakes. Compaq Computer stepped into the market with a "luggable" computer when IBM was struggling to meet demand for its own products. Novell grew rapidly in the networking business at a time when its arch competitor, 3Com, was trying to get into the hardware business and therefore was paying less attention to software. MIPS Computer launched its RISC (Reduced Instruction Set Computing) processors and Sun its SPARC architecture before Intel and Motorola, the two leading companies, could get their similar products to market. As a result, both Sun and MIPS have their established "camps" while Intel and Motorola don't. Dynabook, a company started in Silicon Valley in 1988, backed by $30 million in venture capital, planned to be the first to market an advanced notebook computer. Problems with the design process and the technology delayed the product introduction. By the time the product was ready the market window had passed. The company's technology was sold for a few million dollars.

Time to market is every management's concern. Today technology and information are at parity. Information is so pervasive that as soon as it becomes visible a new technology is available to everyone. Leadership is much shorter-lived in technology businesses than in other industries. Keeping time to market short is demanded of everyone from the board of directors to the engineering managers.

But for the marketer, time to market *acceptance* is the important factor. How long will it take for the product or service to be accepted by the market? Typically, somewhere near the end of the development cycle marketing is brought in to begin the product introduction plan. The pressure is on. Develop a marketing plan, set the pricing, find and sell the first customers, prepare the distribution or sales channels, train the sales force, and develop the advertising and promotions. Then, as quickly as the hustle began, everything is put on hold when the product is delayed.

Product delays can't always be avoided. But many companies underestimate the difficulty of developing new products. They also underestimate the competition's capabilities and how hard it is to gain market acceptance for a new product. Rather than taking a broad view, both start-ups and established companies tend to work on new products with a narrow, internal focus.

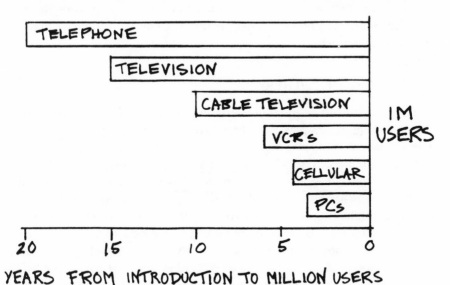

Figure 1.

From the start, companies must think about marketing. Partnering with a few carefully selected prospective customers at the earliest stages of development accomplishes two things: (1) it provides developers with continuing feedback; and (2) it gives the company an opportunity to develop product acceptance even before the product is in the marketplace. GO Corporation, a Silicon Valley start-up, is developing pen-based computing-systems software for the next generation of personal computers. GO refined its systems over a two-year period using feedback from beta users in several businesses, including IBM and a major insurance company. The product was not developed in a vacuum.

Companies must remain close to their customers after product introductions, monitoring their likes, dislikes, and desires. They must see their products' weaknesses from their customers' points of view. But companies must do more than just monitor the response to their products. They also must recognize and correct problems, and look toward future direction, technological advances, extensions to new generations, as well as offer support and service.

The only way to keep time from becoming a nemesis is to build strong relationships with customers and the infrastructure. Company managements must listen to the consumer and stay open to what they hear. They also must keep an eye on competitors. There is nothing paranoid in being paranoid about your competition.

THE RISE OF "OTHER"

There seems to be more of everything today—more cookie companies, more ice cream companies, more brands of clothing, more cars to select from, more personal computer suppliers, more software companies, and more service companies. "Other" owns the leading market share of personal computers, cookies, tires, jeans, beer, and fast foods. Since 1984 American television viewers have been watching "other" more often than the three major networks. Brand names do not hold the lock on consumers they once held. Today, consumers are much more willing to try new things.

Technology advances have enabled companies to cost-effectively design and produce products for a narrower market segment than ever before. Technology is capable of producing an almost infinite variety of product options. In turn, consumers have more choices. Today's consumers have come to value choice; they are not necessarily brand loyal. With so much choice customers have the upper hand. They can be quite fickle about what to purchase and where to purchase it. How then do you create, and sustain, customer loyalty? How do you create a position for yourself and your products that generates customer loyalty?

An article in *Business Week* discussed corporate fads of the '80s. It pointed out that "executives latch on to any management idea that looks like a quick fix." In the '90s we will abandon many of the fads of the '80s and go back to good old "blocking and tackling." Companies will seek to achieve a superior position by building solid relationships with their customers: relationships based on trust, responsiveness, and quality.   *credibility   pg 4.*

It is time for the concept of marketing and positioning to be redefined. The standard promotional concept needs to be abandoned in favor of an approach based on customer and industry infrastructure relationships. To understand the importance of this approach, let's examine some new themes in marketing.

Theme 1: Marketing Is Like Going to the Moon

In fast-changing industries, marketing is somewhat like guiding a rocket ship from Earth to the Moon. The target is always moving. The Moon doesn't stand still, and neither does the market. No two Moon shots are exactly the same. During the flight, you have to keep sensing the environment and making adjustments, altering the course. If the rocket (or the product) is not accurately targeted in motion, it will miss the Moon and may never be heard from again.

Now extend this metaphor further (if you're willing to stretch your imagination a bit). Think of Earth as the company selling the product, and the Moon as the market. Just as Earth and the Moon exert gravitational forces on the rocket and thus influence its course, the company and the market exert their own "gravitational" forces on a product, thereby influencing its course. What are the "gravitational" forces exerted on the product? Let's look at the company's forces first:

The product. Everything begins when the product is conceived. Is it competitive? What is the "whole product?" That is, what other pieces of the solution are needed to make the customer feel completely satisfied? The product is also defined by the company's reputation and the salesperson or distribution channel.

Technology. Does the company have all the technology it needs to develop the product? Is its technology at the cutting edge?

Financial resources. Does the company have enough money for the product development effort? If so, does it also have enough money to support the product with the proper marketing, service, and peripheral products?

Timing. In fast-changing industries the window of opportunity can close quickly. Will the company bring the product to the market at the right time?

Service and support. In the new competitive company these are not "fix-it" operations; they are integral to relationship marketing. Has the company made a service and support commitment?

People. People are the most important ingredient for success. Does the company have top-notch talent in its engineering, marketing, sales, and managerial ranks?

The company can control all these forces, at least to some degree. The most difficult task is being brutally honest about the true competitive capabilities of these resources. In some cases these forces hold the product down. In other cases they give the product a strong lift-off.

The market's gravitational forces influence the product at the other end of its journey. They draw the product in; help position it in the minds of the customers. They help give a product credibility—or rob it of credibility. The market's forces, which follow, are as important to a product's success as the company's forces:

Strategic customer relationships. Are existing customers willing to help define new services or products? Are present customer relationships strong enough to assist with new-product introductions? Companies can form all types of relationships: equity investments, joint development ventures, and marketing and distribution agreements. A company's credibility in a market depends on the relationships it forms. For example, Microsoft's credibility in the software industry shot up sharply when IBM decided to use an operating system from Microsoft for its personal computer. Despite a falling out between the two companies, Microsoft has built so many industry relationships with chip suppliers, computer companies, other software companies, and customers that its position is going to be very difficult to preempt.

Market infrastructure. The infrastructure includes everybody that can influence consumer perception of the product; retailers, distributors, financial analysts, peripheral manufacturers, and others. Support from the infrastructure is critical to success. It has been established that IBM spent $100 million advertising the PCjr. Yet the product failed. It failed because the early users, third-party software suppliers and dealers, found objections to the product. Criticism from the infrastructure killed the product. The same holds true for the Apple LISA, Lotus's Symphony, and New Coke. All were heavily promoted. All failed.

FUD. This stands for Fear, Uncertainty, and Doubt. If customers have fears and doubts about a product, that product won't sell well, no matter how technologically advanced it is. If you plan on challenging a large established company, the distribution channel will be less likely to take on your product for fear of stepping on the toes of the bigger, established supplier.

Competition. The actions of competitors can turn a product into a smash—or a flop. A product might look good at the launch, but a new product, using a new technology, can make it look obsolete overnight. New entrants almost always underestimate the competition.

Adaptation sequence. The market adapts to new technology in stages. First a handful of future-oriented customers, the "Innovators," will try a new technology. Then come the majority of the customers. Finally the "Laggards" adapt to the technology. Where a product falls in this adaptation sequence certainly influences its chance of success. *(See Figure 12.)* ?)15

Social trends. The prevailing views of society can greatly influence how a product performs in the market. In many ways the environment defines the product. For example, in critical social demand areas, such as AIDS or cancer research, a new biotechnology product can create a unique position for the new-drug developer. The hobby or personal computer was an integral part of the countercultural movement of the late '60s and early '70s. The environment helped establish the momentum for the market.

These gravitational forces are always shifting and changing. Nothing in the market is static. Marketeers will succeed only if they constantly evaluate these forces and react to changes in them. Competing in a dynamic market requires a dynamic marketing strategy.

Theme 2: Marketing Is About Market Creation, Not Market Sharing

Most people in marketing have what I call a market-share mentality. They identify established markets, then try to figure out a way to get a piece of those markets. They develop advertising, merchandising, pricing, and distribution strategies geared to winning a piece of the market. This strategy is aimed at winning market share from other companies. Compare this to the Japanese strategy of approaching key industries and segments. They join forces with competitors, acquire technology, invest in user values—quality, design, miniaturization, price, distribution, and service—tying together their investment and market-dominating strategies. Clearly this is a resource battle that only the well-financed, resource-intensive can play. But there is another strategy that allows the risk taker, the innovator, and the smaller player to win. This new strategy is one of new-market creation.

In the mid-1980s Apple was struggling to gain entrance into corporate America. All the pie charts showed IBM dominating the business market for personal computers. I remember one marketing manager presenting a market-share analysis and describing what Apple had to do to win a share of that market. He advised that Apple must look more like IBM, adopting IBM's operating system as well as the same applications software. He concluded that in this manner Apple could achieve 3 to 5 percent of the business market. I got up and drew a circle on the board. "Now," I said, "how can we define the market so we own the whole thing?" Apple had the leading technology in graphics. It also had launched the first laser printer with Adobe software (a strategic relationship). By talking to customers in the newspaper and research businesses, we found an enthusiastic response to the new-found easy way to create graphics. Apple, with the help of software from Adobe and Aldus (Pagemaker), launched a whole new market: desktop publishing. Now Apple's share dominated the pie chart and IBM had an insignificant portion. Not only was the strategy a success but desktop publishing became Apple's Trojan Horse into corporate America. Every major corporation has a publishing operation and the superb graphics produced by the Macintosh soon became visible to every executive in those corporations.

Market-sharing and market-creating strategies require very different sorts of thinking. Market-sharing strategies are common in mature consumer-goods industries like soft drinks and rental cars. The emphasis is on advertising, promotion, pricing, and distribution. Customers are interested primarily in price and availability. Product differentiation is largely cosmetic. Suppliers with the best financial resources are likely to win the largest market share.

That is, they win market share, but often at the expense of profitability. Apple's market share makes up about 10 percent of the personal computer industry, yet it is one of the top three or four most profitable computer companies in the United States. In his book *Competitive Strategies, Techniques for Analyzing Industries and Competitors*, Michael Porter points out that "there is no single relationship between profitability and market share, unless one conveniently defines the market so that focused or differentiated firms are assigned high market shares in some narrowly defined industries, and the industry definitions of cost leadership firms are altered to stay broad (they must because cost leaders don't often have the largest share of every submarket). Even shifting industry definitions can't explain the high returns of firms who have achieved differentiation industry-wide and hold market shares below that of the industry leader."

A market-creating strategy is a differentiation strategy. Functioning within this strategy, managers think like entrepreneurs. They are challenged to create new ideas. The emphasis is on applying technology, educating the market, developing relationships with the industry infrastructure, and creating new standards. The company with the greatest innovation and creativity is likely to differentiate itself and win the desired market.

Traditional market-share strategies don't work well in emerging markets. Most new markets are quite small to begin with. If companies think only about sharing the markets, they never will get involved in emerging businesses. They'll take a look at a business, decide that the "pie" is too small, and move on to other possibilities.

That is exactly what happened in the personal computer business. In the mid-1970s dozens of major companies investigated the market for inexpensive computers. At the time, these computers were used primarily by hobbyists—enthusiasts who enjoyed tinkering with the machines. There aren't many hobbyists of this sort in the United States, so most computer companies decided that the market was too small for them to enter. But a few companies, companies such as Apple and Tandy, looked at the business with a market-creation mentality. They looked beyond the hobbyists and saw that small businessmen and professionals might eventually use the machines, if only the machines were designed and marketed a bit differently. Rather than focusing on what was, they focused on what might be. They saw the possibility of creating a growing market, an expanding pie, and they set out to make that happen.

In creating new markets, marketeers face many obstacles. First of all, they can't rely as much on analogies and case studies for guidance. When products or services are radically new and different, past products or services don't provide much of a guide. The personal computer, for example, had no good analogies. Clearly, personal computers aren't at all like large mainframe computers. They sell at very different price points to very different people. Some people have compared personal computers to stereos. But stereos are much less complicated to use than computers. People don't spend hours learning how to use a stereo. Most people aren't as intimidated by stereos as they are by computers.

For cases like this, marketeers must break new ground. They must be willing to experiment and take risks. They must try new things and be open to new ideas. Creativity is the key to success in new markets. In mature markets, marketing is like a handball game—a confined environment with few players. But in emerging markets, marketing is more like a soccer game—a wide-open field with lots of players, lots of possibilities, and lots of options.

In the early days of personal computers, no one knew how to distribute the new machines. The traditional method for selling computers—through direct-sales forces—was simply too expensive. A person with a market-share mentality might have given up. But a few innovative people persevered. Some tried direct mail, others tried selling computers door to door. Still others opened specialty computer stores. Within a few years there were thousands of computer stores scattered across the country. New channels can be created or old channels used in new ways. Dell Computer astonished the industry by building a high-quality company based on mail-order sales. The company's quality is outstanding and its service and support are so exceptional that they have virtually eliminated consumers' fear of buying a complex product sight unseen.

To develop new markets, it is essential for companies to take the time to educate customers. When microprocessors were first introduced, in the early 1970s, few customers recognized the value of the new chips. People are resistant to change, and the idea of programmable chips was foreign. Even many engineers believed that the microprocessor was a marketing gimmick.

So Intel, the first company to market microprocessors, had to do a massive education job. It ran advertisements filled with suggested applications for the new product. It distributed booklets containing descriptions of actual applications, from electronic games to blood analyzers, from milking machines to satellites. Most important, Intel ran seminars for potential corporate customers. In the first few years Intel ran hundreds of these seminars worldwide. At each seminar, Intel first presented a corporate overview, usually given by a top company executive. Next, an Intel marketing manager would give a presentation on the marketing value of microprocessor-based products. Finally, Intel engineers would describe the technical details of the microprocessors. Most of the early customers ordered only a few microprocessor chips. But as the education campaign continued, Intel attracted more and more high-volume users.

Intel listened and learned from its early customers. The first microprocessor was developed in response to a Japanese customer's request for help with solving a problem. As Intel worked with early customers it discovered what was good and what was bad about the technology. Development systems and tools were made available; peripheral chips and software, as well as new generations of more powerful chips, were brought to market; all because the early users demanded it. By aggressively pursuing both the technology and the customer, Intel established itself as the industry leader and set the standard for all to follow.

A market-creating mentality also requires a different view of industry standards. Companies must think about creating new standards rather than following existing standards. That involves greater risk, but the payoffs can be much higher.

Apple took this route when it developed its Macintosh computer. Rather than striving to produce a clone of the popular IBM personal computer, Apple wanted to develop a new computer that was radically easier to use. To do that, Apple decided to establish its own operating systems rather than be bound by the limitations of the industry standard MS-DOS. The risks were great, but to have done otherwise would have left Apple completely at the mercy of IBM's control over that standard. Clearly its choice of strategy has allowed Apple to differentiate itself and achieve significant profitability.

Theme 3: Marketing Is About Process, Not Promotional Tactics

In 1983 IBM introduced the PCjr with a multimillion dollar promotional campaign. It ran commercials on television and placed advertisements in dozens of magazines. But with all that promotion the PCjr still didn't sell well. The PCjr's toy-like keyboard and lack of software compatibility were part of the problem. But equally important was the fact that dealers were not excited by the machine. IBM had worried so much about winning the minds of customers that they had never won the hearts of the industry infrastructure.

In mid-1984 IBM tried to fix the problem. One step was to redesign the keyboard. But equally important, IBM invited all its retailers to a huge meeting in Dallas. Company officials gave the retailers technical information, sales advice, and a big party. They listened to retailers' questions and complaints. The retailers left the meeting with a new attitude toward the PCjr, but the damage had already been done. Despite tens of millions of dollars spent on advertising and promotion, the PCjr was not accepted by the infrastructure. So, in 1985, IBM announced that it would no longer manufacture the PCjr.

The moral of the story? Advertising and promotion are only a small part of marketing strategy. Advertising can reinforce positions in the market, but it can't create positions.

To build lasting positions in the market, companies must first build strong relationships. They must build relationships with customers, suppliers, distributors, resellers, industry influencers, and members of the financial community. They must take advantage of the industry infrastructure—the key people and companies that make the industry tick.

I like to draw a distinction between marketing-driven approaches and market-driven approaches. The two approaches are very different. Marketing-driven approaches are based on using a bag of marketing tricks, gimmicks, and promotions in order to capture the mind of the consumer. Market-driven approaches involve dialogue between the company and the customer, and between the company and the market. Products, services, and approaches to the market are altered, modified, changed, and often created by such dialogue.

Traditional consumer-marketing or commodity-goods producers use the marketing-driven approach. They often act as if some set of words and pictures will suddenly capture the customers' loyalty. That is unlikely in today's environment of diversity and choice.

Technology products or any complex or high-risk products or services require acceptance more than awareness. References and relationships play a major role in the purchase decisions about these products. No one buys a personal computer, mainframe computer, data-base management system, digital switch, accounting software, microprocessor, or expensive sports car without first obtaining favorable references from both users and other important third parties within the market-making structure. In like manner, no one picks a family physician or lawyer from the Yellow Pages. Marketing complex products is much like marketing services. Such marketing must be based on knowledge, information, trust, relationships, word-of-mouth references, and leadership.

New technologies seem to emerge every day. Regulatory changes, like the breakup of AT&T, restructure entire industries. Abundant venture capital, particularly here in the United States, encourages and supports innovation and change. Much of this change is driven by the computer. More than 50,000 computers are sold in the United States alone every day. These computers improve productivity; solve complex problems; and create new processes, products, and services. Every business is affected by new computer technologies. Computer-based industries produce a staggering diversity of goods and services. Product life cycles can be as short as a few years, and new discoveries turn industries upside down. Every industry and every business is affected. Little wonder that customers in these industries need security and reassurance.

Buying a $1 tube of toothpaste doesn't involve much risk. But buying a $25,000 computer system that will be at the heart of a business is a major risk. Customers are filled with worries: If the computer breaks down, will my business grind to a halt? Will the manufacturer provide prompt, quality service? Will my new computer be obsolete in a year? If so, will the manufacturer introduce new, up-to-date models that preserve my investments of the past? As my business grows and I want to upgrade my equipment, will the manufacturer offer a smooth path to larger computers? Will other companies provide the software and peripherals I need for the computer? These all are legitimate fears.

Running more advertisements will not ease these fears. People are deluged with so much product information these days that information has become disposable. With more than 400 different personal computers on the market, people aren't going to decide which one to buy on the basis of advertisements. They are going to rely on the advice they get from resellers, consultants, and friends.

There's an old Texas saying about a cowboy who was "all hat and no cattle." That is, he was all show and no substance. Technology-based companies can't build an image that way. They have to have the cattle. If they don't, then despite the strength of their advertising and promotion, new technological developments will leave them in the dust.

For that reason, companies in technology-based businesses must use market-driven, not marketing-driven, approaches. They should concentrate on substance before image, for it is substance that supports the image. Rather than try to reach customers through a Madison Avenue ad campaign, they must create a dialogue with customers. They also must support, educate, and develop the infrastructure or market-support structures. It is solid relationships with members of this infrastructure that help to support and establish a company's products. Jim Morgan, CEO of Applied Materials, a manufacturer of semiconductor equipment, once noted: "Image is a collection of things that we do in the marketplace." Morgan is not a marketing man, but his instincts are right on the money. If a company produces a solid product and builds relationships properly, its image will take care of itself.

Theme 4: Marketing Is Qualitative, Not Quantitative

Businessmen love numbers. Numbers make them feel secure. But in emerging markets, numbers are rarely reliable. Marketeers who rely on numbers are unlikely to succeed.

In many cases, quantitative analyses use the past to predict the future. But we live in an era when the future almost never resembles the past. It is extremely difficult to take the pace of technology into account. Extrapolating today's trends into the future is futile. No one can predict the future. It strikes me funny that we laugh at Jeanne Dixon and other clairvoyants who predict events each New Year, yet businesses routinely spend untold hours trying to predict events five and ten years out.

Companies have run into this problem since the dawn of the computer age. In the 1940s computer companies made estimates of the total world market for computers. They calculated the total market at several dozen computers. That's all. Several dozen computers for the whole world. They simply didn't anticipate the proliferation of new applications, or the sharp decline in computer prices.

Mitch Kapor, developer of the incredibly successful software program Lotus 1-2-3, ran into a similar problem when he was developing his original business plan for the integrated software package. He developed the business plan in the late 1970s for a course at the Massachusetts Institute of Technology's Sloan School. Kapor received a grade of B, rather than A, for his project, in part because he included no statistical market surveys.

What would Kapor have found if he had done a statistical survey? He probably would have found no demand for his product. After all, hardly any large corporations had personal computers in the late 1970s. But Kapor had a sense of the market. He knew that corporations would eventually buy personal computers, and then they would want his software.

Kapor used what I call a *qualitative* approach to the market. He had talked to people in the market. He understood their needs on a human level. A qualitative approach to the market goes beyond the numbers to explore the trends and perceptions that create the numbers. It looks at customer attitudes and personal relationships. Only by understanding the market in a qualitative way can marketeers hope to create the future.

Start-up companies with truly innovative, market-creating products often have a difficult time raising money because they can't accurately size the potential market. Venture capitalists are not always very venturesome if they can't put numbers on the market. Any forecasts in such circumstances are only indicators. Too many factors influence the product adoption rate. Besides, markets don't exist independent of the products or the people who create them.

I recently looked at the 1978 prognostications of half a dozen research firms predicting the worldwide 1985 market for personal computers. The most optimistic forecast showed that the market would be about $2 billion. In fact, the market exceeded $25 billion. But those forecasts didn't deter Steve Jobs. Nor did they change Ben Rosen's and Rod Canion's minds about starting Compaq Computer in 1981. These men had decided to make the market happen.

Even when information is accurate, judgments made about it are often wrong. Several years ago I was looking for a valid way to obtain objective information. I was referred to a man who had once been head of information systems at the CIA. The CIA had to have sophisticated ways of gathering and interpreting information, I thought. This man told me that the information-gathering techniques were rather conventional: use of satellite photographs, agents, newspapers and magazines, technical reports, and other obvious means. But then he said that when the same information was given to two different administrations in the White House, he got back two different interpretations. We can have reams of data on the competition without being more competitive and we can have lots of information about our customers without being customer sensitive. Shortly after the Columbia shuttle disaster I heard a NASA official say on the MacNeil Lehrer News Hour, "We were so overwhelmed with data that we lost sight of common sense." The biggest problem with market data, therefore, is not the data itself but rather the judgment made about the data. Historically, businesses have used abstract market data, drawing conclusions without ever touching living, breathing customers.

Bare statistics miss the nuances of the market. A survey might show that 60% of all customers use a company's product. But a qualitative approach might reveal that the customers are unhappy with the company's service, and that many are considering switching to a competitor. A survey might show that if a certain product were available with certain characteristics,

THE FAR SIDE   By GARY LARSON

"Bob and Ruth! Come on in .... Have you met Russell and Bill, our 1.5 children?"

Figure 2.

people would buy it. But there is no way of knowing that for sure. There is infinite demand for the unavailable.

Robert Kennedy once observed, in reference to measurements of the Gross National Product, that we can measure everything except those things that are worth measuring. Yet as companies grow they tend to rely heavily on quantitative techniques. They lock themselves into numbers and abstractions. They end up with products that don't match the needs of the market nearly as well as entrepreneurs' products. They have squeezed creativity out of the system.

We need more companies to act like entrepreneurs. Successful entrepreneurs don't worry about statistical market projections. They don't care whether projections show a $5 million market or a $500 million market. They plan strategy in a qualitative way. They simply take good ideas, develop them into products, then constantly adjust the products to market needs. They create their own opportunities by pushing as hard and as fast as they can.

A qualitative approach is important in sales as well. Many technology-based companies try to sell their products based on quantitative specifications. They boast that their product has an access time of so many nanoseconds, or a capacity of so many kilobytes. But customers tend to make their decisions on more qualitative factors, such as leadership, service and reliability, and reputation. If a company can establish credibility with key people in an industry, it is likely to succeed, even if its product is technically weaker than the competitive product.

To take a qualitative approach to marketing, managers must understand what I call the market environment. The environment includes all the gravitational forces mentioned in the Earth–Moon analogy—things like social trends, relationships, and competition. Each of these forces *(see Figure 3)* influences the way customers perceive the product.

Quantitative approaches to marketing often ignore the environment. They view products as isolated objects that can be defined by statistics and specifications. But products in the real world are not isolated objects. They exist only in the context of their environment. Qualitative approaches to marketing use the environment as a guide to help companies understand products and markets.

The U.S. semiconductor industry has gathered statistical information for as long as I can remember. It could tell you how many products of which kind in what packages existed and to what industry they were being sold. It could forecast the growth and size of each market segment. But in 1979 most semiconductor companies were taken completely by surprise when quality became an issue. Key users gave testimony to the superior quality of Japanese products. If the members of the semiconductor industry had built strong relationships with their customers, Japanese encroachment into the memory market on the basis of quality and price might never have occurred. Quality has since become an essential marketing and environmental element.

The environment acts as a lens through which the customer views the product. As the environment changes, so does the public perception of the product—even if the product itself has not changed at all. As technology advances, products that were once perceived as being on the cutting edge begin to look mundane. As prices drop, products that once seemed cheap begin to look expensive. As new issues arise they become part of the environment.

To market a product effectively, marketing managers must understand the workings of the environment. Managers must be sensitive to trends and consumer perceptions. They must understand how various forces in the environment interact with one another, and they must be alert for changes in these forces. In effect, they must see their products as customers see them—through the lens of the environment.

Take the personal computer industry as an example. In 1977, when the Apple II computer was launched, the personal computer industry was still in its infancy. *(See Figure 13.)* It had little competition and lots of opportunity. Hobbyists were the primary customers, and retail computer stores had just begun to open.

The environment was perfect for Apple. Other industries were full of bad news. Japanese companies were beginning to dominate the automotive and consumer-electronics industries, and they were even making inroads in the semiconductor industry. The American public, and American journalists, were eager for some good business news.

Apple capitalized on this environment. It presented itself as a symbol of hope for the future. It was a bright spot in an otherwise dull and depressing business environment. America was beginning to look upon entrepreneurs as the saviors of American capitalism. The rags-to-riches story of Apple's dynamic founders, Steve Jobs and Steve Wozniak, spoke to the environment. Other personal computer companies emphasized the technical specifications of their products, and made elaborate presentations on the technical differences between brands. But Apple recognized that the environment involved more than hardware. The personal computer industry was in its infancy; there was room for everybody. The main challenge was to attract new types of users. So, to meet that challenge, Apple stressed the fun and potential of the new technology. In short, the 1977 environment for personal computers was nonthreatening and curious. Apple took a qualitative approach to marketing, and that decision turned the Apple II into a big winner.

Marketing is a way of doing business. A well-known chief executive once said, "Marketing is too important to leave to the marketing people."

Businesses are not going to do away with the marketing department anytime soon. But we must begin to include the whole organization—the entire company, from the chairman of the board to the telephone operator—in the marketing business. All employees need to be in the business of building customer relationships.

Theme 5: Marketing Is Everybody's Job

Marketing is building and sustaining customer and infrastructure relationships. It is the integration of customers into the company's design, development, manufacturing, and sales processes. In order to achieve a distinctive position in any industry, the whole company must think about being in the marketing business. Engineering or development people can have a better sense of what to build, what to change, and how to fit the product into the customer's existing systems if they act as marketing representatives. It's no secret that today's manufacturing operations are in the marketing business.

In an effort to convince a major value-added reseller to switch from a competitor, Silicon Valley graphics-systems company Radius brought the reseller's management to the see its production operations. Radius had invested early in automated production and testing equipment. Its entire operation was focused on ensuring quality, reliability, and timely delivery. The reseller was so impressed by the sophisticated operation that it changed suppliers.

Convex Computer, a supercomputer company in Texas, is up against formidable competition from Cray, IBM, and Digital Equipment as well as from the Japanese. Hanging high on the walls at its manufacturing operations are flags of the many countries to which Convex computers are shipped. Graphs and charts showing declines in defects, improvements in reliability, and the various quality ratings are displayed throughout the manufacturing area. The operation is clean, well organized, and efficient. Robots, guided by cameras, insert chips into sockets on computer boards. Terry Rock, the head of manufacturing, can discuss detailed competitive advantages of the Convex machines, the dollar function values on international exchanges, export license issues by country, Baldridge Award requirements, cost issues, issues regarding installation of software and hardware, examples of satisfied customers and their applications, and other topics customers bring up. Terry is an important part of the marketing operation. Many Convex customers are sold on the company by visiting its manufacturing operations.

Probably one of the most significant customer-driven innovations of the past quarter century was the microprocessor. In 1969 Bob Noyce, founder and then president of Intel Corporation, and engineer Ted Hoff were visiting Busicom, a Japanese calculator company. The problem facing the Busicom engineers was to reduce costs by designing a calculator with as few chips as possible. On the way back to California, Hoff said he thought he could get the calculator functions down to a few chips, and eventually maybe even just one. The result was the first commercial microprocessor. In solving a problem for his customer, Hoff created a new industry. In this situation he was as much a marketer as an engineer.

These five cornerstones of the new marketing should serve as guides throughout all parts of the positioning process. Once again, they are:

1. Marketing is like going to the moon.

2. Marketing is about market creation, not market sharing.

3. Marketing is about process, not promotional tactics.

4. Marketing is qualitative, not quantitative.

5. Marketing is everybody's job.

Now let's go on to discuss the positioning process. First we'll look at the role the customer plays in this process. Then, in the following several chapters, we'll look at product positioning, market positioning, and corporate positioning—the three stages in what I call the dynamic positioning process—to show how companies can put the new marketing to use in building a successful marketing strategy.

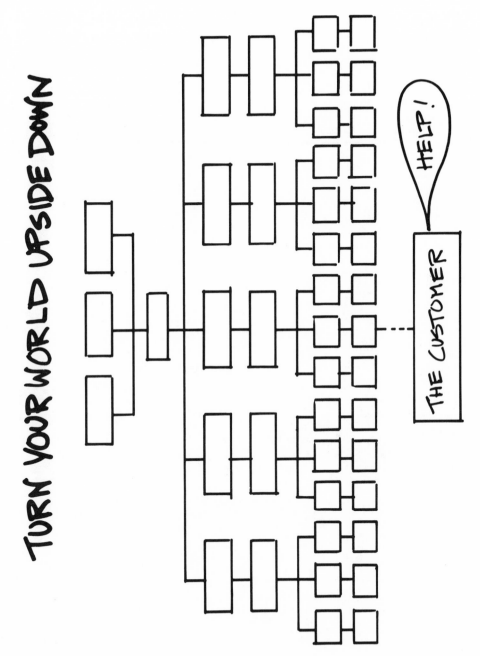

Figure 3.

Listening to customers *must* become everyone's business.

Tom Peters

Chapter 3 | Relationship Marketing: It All Starts with the Customer

Positioning begins with the customer. Customers think about products and companies in relation to other products and companies. What really matters is how existing and potential customers think about a company in relation to its competitors. Customers set up a hierarchy of values, wants, and needs based on empirical data, opinions, word-of-mouth references, and previous experiences with products and services. They use that information to make purchasing decisions.

Market leadership can catch a consumer's attention and can be an important factor for the customer to consider, but there is more than one leader in almost every market segment today. Most often several companies share the top position, one being the technical leader, one the market leader, another the pricing leader, and still another the challenging upstart. If I say "color television," people probably think of Sony, Mitsubishi, or Hitachi. If I say "on-time delivery," they no doubt think of Federal Express. This is one of the few companies that has been able to develop a stand-alone, distinctive position. If I ask for names of "personal computers," people will come back with several names, including Apple, IBM, Microsoft, Lotus, Compaq, Toshiba, and Dell Computer—all companies recognized as having distinct market positions. If I qualify further and say "innovative computers," chances are people will think of Apple. If I say "3-D graphic workstations," the response would be Silicon Graphics. Now upstarts, GO Corporation and Momenta, are challenging industry mainstays by developing what many believe to be the new-wave, pen-based computer software.

As the markets have fragmented, more players are able to achieve a recognized position for themselves even without the resources of Sony, Apple, or Federal Express. Remember, even the giants were once small and unknown. Multiple positioning opportunities exist within every industry. A distinct, market position is attainable even in a crowded marketplace. Just as Ben & Jerry's ice cream and Dell Computer have succeeded in differentiating themselves in markets full of giants, companies today have many opportunities to build a distinctive customer following.

Many companies can establish unique positions in the marketplace for themselves, their products, and their services. This capability is a powerful force in marketing. Indeed, at the heart of every good marketing strategy is a good positioning strategy. But positioning is not so much what you say about your products or company as much as it is what your customers say about you. It is not what you say to your customers but rather what you do with your customers that creates your industry position. Differentiation, from the customer's viewpoint, is not something that is product or service related as much as it is related to the way you do business. In the age of information, it is no longer possible to manufacture an image. The distinction between perception and reality is getting finer. Further, in a world where customers have so many choices they can be fickle. This means modern marketing is a battle for customer loyalty. Positioning must involve more than simple awareness of a hierarchy of brands and company names. It demands a special relationship with the customer and infrastructure of the marketplace. If companies are to develop a marketing style suited to this era of rapid change, they must start with a new approach to positioning.

Positioning is vital to success in marketing. All of marketing's components—competitive strategy, pricing, packaging, distribution, service, support, communications—are interrelated in positioning strategy. If a company's products are positioned poorly, the problem may be in the design, manufacturing, or marketing of those products.

How can a company establish a strong position in today's fast-changing markets? It isn't easy. Traditional positioning strategies are inadequate because they don't consider technology and change to be significant factors and, more importantly, because building and sustaining customer relationships is never their central aim. They assume that a static, impersonal customer and marketplace—that is, a marketplace where technologies, products, and customer perceptions change very slowly—exists. Marketeers need a new model of positioning for use in today's markets. They need what I call dynamic positioning.

Dynamic positioning strategies are very different from traditional strategies. With the traditional model, a company first decides how it wants to be positioned. The company might want to be perceived as the industry's low-price company or perhaps as its premium-quality company. Next, it comes up with a slogan that summarizes the desired message. Finally, it simply spends money on advertising and other promotions until the slogan achieves broad recognition. This method is based on manipulation of the customer's mind; on using a bag of marketing tricks to entice the customer into awareness of a company's desired position. Such positioning theory is company centered rather than customer centered. In today's market it is static.

The Avis–Hertz rivalry is a classic example of traditional positioning. Avis decided on a position: the hardworking runner-up in the industry. Then it came up with a slogan: "We try harder." Finally, it advertised like crazy, until people began to believe that Avis really did try harder. This approach worked because the rental-car business is in a rather static market. Neither the cars nor the service changes much from year to year. If you rent a car, you rent a car. Static-market companies used to differentiate themselves simply by advertising, and by offering discount rates and free gifts. Even in these businesses, however, the situation is changing. Today such companies must differentiate themselves not by being the biggest, but by providing the best service.

re: new Hertz adds

The situation is quite different in complex, fast-changing industries. In these industries, radical changes occur every day. Products evolve, markets change, and new technologies emerge. The competition changes as well. New companies, and established companies from other industries, constantly are trying to grab a piece of the market. All these changes can influence positioning in the marketplace.

Standard approaches to positioning don't work in the new customer-centered environment of the 1990s. A company that is Number 1 today has no guarantee of being Number 1 tomorrow. New technologies can turn a seemingly solid position into a fragile one almost overnight. No amount of advertising can prevent that from happening. Even with the best of slogans, a company can lose its position in the market. Unisys spent millions of dollars building awareness (The Power of ^{2}), but its position continues to deteriorate in the marketplace.

Market awareness is no longer enough to carry a company to success. An associate of mine once did an analysis of the Federal Energy Administration's slogan "Don't be fuelish," which was promoted during the 1973 oil crisis. About 80% of all people surveyed were familiar with the slogan. But energy consumption remained at an all-time high. The slogan was good, but public behavior was unaffected. However, the law reducing the speed limit to 55 mph did have a dramatic effect on fuel consumption. Action, not words, make things happen. We all are *aware* of many products and services that we would never consider buying.

To survive in dynamic marketplaces, companies clearly need to establish strategies that can survive the turbulent changes in the market environment. They must build strong foundations that will not be blown away in the storm. They won't do that by focusing on promotions and advertising. Rather, they need to gain an understanding of the market structure. Then they must develop relationships with suppliers and distributors, investors, customers, and other key companies and people in the market. Those relationships are more important than low prices, flashy promotions, or even advanced technology. The feedback loop is central to these sorts of relationships. Customers and others influence changes in products and services through their participation in the relationship. Changes in the market environment can quickly alter prices and technologies, but close relationships can last a lifetime.

With this approach, positioning evolves gradually. Company or product positioning is somewhat like development of a person's personality. Babies have no real personality when they are born, but they gradually gain characteristics as they grow. They are influenced by their parents, then by their friends, then by education. Their personalities alter, grow, and adapt to the environment and relationships surrounding them.

Similarly, a product or company in its infancy has no real meaning. But it acquires meaning from its environment, and it changes as the environment changes. As a company evolves it is still the same company, just as a growing child is still the same child. But personality and positioning are always changing.

Unlike traditional positioning, dynamic positioning is a multidimensional process. It involves three interlocking stages: product positioning, market positioning, and corporate positioning. *(See Figure 4.)* These three stages interact with one another in subtle but important ways. Each stage builds on the others and influences the others. Pieced together properly, the stages create a whole that is much bigger than its parts. But if any one of them is flawed, then the whole positioning process will falter.

In the first stage—product positioning—a company must determine how it wants its product to fit into the competitive market. Should it build a reputation for low cost? High quality? Advanced technology? How should it segment its markets? Who should be targeted as the first users? I always advise companies to pay special attention to intangible positioning factors, such as technology leadership and product quality. Intangible factors are based on customer perceptions, not on raw statistics and numbers. Marketing is not a rational process. Low prices and top product specifications don't always

DYNAMIC POSITIONING

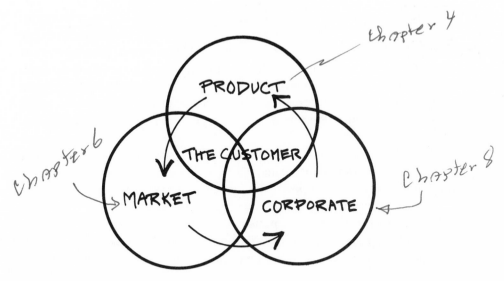

Figure 4.

win sales. Rather, it is intangible factors that are the keys to building customer relationships and gaining strong product positioning.

In the second stage of the positioning process—market positioning—the product must gain recognition in the market. It has to establish credibility with customers. The marketplace must perceive the product as a winner. In his book *Competitive Strategies, Techniques for Analyzing Industries and Competitors*, author Michael Porter says, "Every industry has an underlying structure, or a set of fundamental economic and technical characteristics, that gives rise to competitive forces. The strategist, wanting to position his company to cope best with its industry environment or to influence that environment in the company's favor, must learn what makes the environment tick."

To gain a strong market position companies need to understand the participants in the industry infrastructure: early customer advocates, the reseller networks, distributors, and third-party suppliers, as well as analysts, industry "luminaries," and journalists, who control the flow of industry information and opinion. Companies should identify and work closely with key members of the industry infrastructure. I believe that 10 percent of the people in an industry strongly influence the other 90 percent. If a company can win the hearts and minds of the most important 10 percent, its market positioning is assured.

In corporate positioning, the final stage in the process, companies must position not their products, but themselves. This is done primarily through financial success. When companies are profitable, many of their mistakes are forgiven, if not forgotten. But when a company's profits slip, their position is tarnished. Customers are reluctant to buy products, particularly expensive or complex ones, from companies in financial trouble. If a company finds itself in this situation, it must start over at product positioning and rebuild its position in the market.

This three-stage positioning process must be central to all business activity. It is not a promotional game. It is a fundamental part of business planning, and it must be supported by managers across the corporation.

Dynamic positioning draws a common thread through all parts of the company, then connects these parts to the marketplace. Dynamic positioning must be a part of the total organization, and in turn, it will have a major influence on every aspect of the organization, including:

Corporate image. Positioning can influence the attitudes of staff members. People enjoy working for a company they can identify with, especially if the company is recognized as a leader. Certainly, recruiting is easier for recognized leaders. Positioning can also influence the company's relationship with the financial community. Wall Street likes companies with a clear vision of their role in the market.

Product planning. Product planners are engaged in a constant battle with change. Regular positioning analysis can provide direction for overcoming weaknesses or creating barriers to competition. Product planning is best done in a dialogue with the customer and with a full understanding of the competitive environment. Product planners must move away from traditional marketing techniques that were designed to gain market share and move toward new approaches designed to create entirely new markets.

Marketing. Marketing mostly involves building relationships and, through them, guiding the company's future. Marketing managers must be the integrators of the company's positioning. They must project the position to the market through education and by establishing relationships with members of the infrastructure. Strong positioning allows a company to establish relationships with strong partners. These relationships, in turn, make the company's positioning even stronger.

Financial health. Positioning and financial strength build on one another. A well-positioned company can raise new funds more easily. Conversely, a financially strong company has a much easier time positioning its products in the market.

THE DEATH OF A SALESMAN

Dynamic positioning is particularly important in sales, because successful salespeople develop customer relationships, not just orders. How has the job of selling changed with the times and new technology?

The short answer is that in some cases, it hasn't. Door-to-door selling is still the most lucrative way of getting products like Avon, Girl Scout cookies, and Fuller brushes to customers. However, sales presentations supported by complex computer graphics work best for selling software to engineers. In the high-tech arena, the salesperson's newest trick is a kind of disappearing act. Company and customer are working so closely together that selling, no longer a discrete function, is subsumed in problem solving. As a recent *Wall Street Journal* article noted, "It's a rare company whose marketing people don't say. . ., 'Our competitors sell boxes, but we sell solutions.' "

Selling is a disparate task, tailored to the times and to the vicissitudes of different types of goods and services. Over the last century, salesmen have peddled their wares on foot, by horse-drawn wagon, railway, telegraph, automobile, airplane, telephone, radio, television, and now by computer and satellite. Because sales are so directly linked to profits, new tools and techniques for winning new business are always welcome.

Today, a "salesman" can be anything from a decidedly non-human computerized phone connection to a sophisticated purveyor of information whose consultation and services can bring added profits to your business.

High-technology salespeople draw on a broader spectrum of skills than did their predecessors, and these new skills boost their image in the business world. They are seen not as peddlers but as knowledgeable professionals. Armed with higher education, sales training, persuasive persistence, and a constant stream of updated information from computers, high-technology salespeople have become the Great Differentiators for their products and companies.

Given the confusing array of reliable, mostly low-cost computer products, how is a consumer to choose among products that seem much the same? I often describe a computer as being 80 to 85 percent service, which is another way of saying that the salesperson is an integral, intangible part of the product he sells. He provides the vital extra ingredients that differentiate his product from all the others—application knowledge, system analysis and integration, service support, and training. A salesperson's knowledge about his products and how they fit into his clients' often highly volatile business environment can mean the difference between both companies' success or failure.

High technology can enhance the relationship with customers by responding immediately to their needs. Increasingly, salespeople arrive at their prospects' firms with portable computers. At Godiva Chocolates, salespeople use computers in the field to handle order-taking, budgeting, forecasting, note writing, and electronic-mail communications. During a sales call with a department store candy buyer, the Godiva salesperson enters the order. The computer calculates the entire cost, including discount. In an electronic mail message, the salesperson notes that the order is entitled to a special rate during a seasonal promotion. She plugs the laptop into her customer's phone, sends out the order, and receives a delivery date. She can also check her mail. The factory, meanwhile, receives up-to-the-minute information on what needs to be produced.

By giving its sales force laptops and thus access to more product information, Godiva management has given its staff more control. Released from mundane paperwork, salespeople are thought to be more likely to come up with creative selling strategies. It's "like working with a race-car driver," says Godiva president Thomas Fey. "You can't tell him exactly what to do, but we can provide the car and help him maximize his use of the vehicle."

Marketing and sales costs average 15 to 34 percent of total corporate costs, making them a fat target for efficiency efforts. Computer technology specifically designed to automate sales and marketing functions undeniably improves bottom-line profits. A recent Harvard Business Review article compared sales figures from a big and a small company after they had installed marketing and sales productivity computer systems. The big company, a $7 billion electronics firm, increased its sales by 33 percent and gained 31 percent in productivity. Less drudge work and increased productivity created the potential for greater personal earnings. As a result, sales force attrition dropped 40 percent. At the smaller company, an $8 million printing concern, $80,000 invested in telemarketing software and a minicomputer returned a 25 percent sales increase. Both companies recouped their initial investments in less than a year.

New technology can also eliminate any need for salespeople. Howard Anderson, founder of the Yankee Group, the international marketing firm based in Boston, predicts a coming trend in selling called "integrated-customers," meaning that customers and suppliers will be linked with common databases. In Anderson's hypothetical example, Budweiser builds a new brewery in New Hampshire and, by contract, American Can builds a can-making plant next door that is synchronized with Budweiser's needs. Information comes in via an electronic data interchange (EDI) showing that Bud Lite sales are running 27 percent ahead of budget. This data has importance for everyone in the integrated system. It signals the brewery that it will need more water for future production. It tells the sales force that a market shift may be beginning. It tells American Can to buy more steel from U.S. Steel to make more Bud Lite cans. For Budweiser's marketing department, the data may be read as a signal for a new brand, October Fest Lite. Marketing sends an EDI message to American Can's graphic arts department, which immediately gets to work on a label to be embossed on aluminum cans to be ready three weeks hence. Prior to the integrated scheme, producing a new label could take up to six months. The more closely integrated the two marketers become, the faster they can act together for their mutual benefit.

Radius, Inc., which makes full-page on-screen displays, provides another example of how technology could eliminate the need for a traditional salesperson. The software maker's product allows a user to see a whole page of information on-screen, which is useful in desk top publishing and in general business. Radius TV is a product enabling users to open a video window on their computers that displays information taken directly from TV signals. Say an engineer in New York is working with a counterpart in Silicon Valley. Through the video window, the two have a private network via satellite by which they can tinker with product design, draw up blueprints, do calculations, and speak to each other—in short, anything they could do in person short of shaking hands. Farther down the road, the product can be demonstrated to potential customers simultaneously at different locations. The product is demonstrated, even modified, through a reciprocal process. Because producer and client are working together, there's no need for a salesperson. Consumers will become their own travel agents, drawing on information from broadcast or stored video images and calling up schedules and prices simultaneously on their home computer.

Will salespeople become obsolete? As the automobile and telephone radically changed the sales function, so will telecommunications and the new computing environment. The salesperson as a "convincer" or "closer" of orders will go the way of the slide rule and instead become synonymous with "service." The salesperson will carry information, education, training, detailed design, quality, and reliability information. He or she will be the link between the product design and factory and the customer. Today, we are still in the primitive stages of automating the sales and marketing functions. A high level of on-line, interactive exchange between the producer and the consumer is yet to come. I once gave a speech on this subject to an audience that included a Toyota factory manager. I said that one day we will go into an automobile showroom and sit at a computer terminal and design our own car. This seemed like a radical idea to most of the audience—except the Toyota manager. He told me afterwards that the technology is available to have direct customer designing and ordering today. The only obstacle is maintaining and financing the large inventory of parts necessary for the service to be acceptable.

Design and manufacturing technologies will advance to "real-time" processes, which will place increasing demand on the marketing and sales functions to eliminate the gap from design to consumption. I can foresee "marketing workstations" just as we have engineering and design workstations today. The marketing workstation will draw on information from a network of databases containing vast arrays of graphic, video, audio, and numeric information. The marketeer will be able to look through windows on the workstation and see, manipulate, simulate, bounce concepts off others on the network in distance cities, make changes, write production orders for product designs and packaging concepts, and obtain costs, timetables, and distribution schedules. The marketeer will play or simulate both designer and consumer as children play Nintendo today. He or she will request and receive information on historic sales and cost figures, competitive trends, and consumer patterns using UPC information; design and create advertisements and promotions; evaluate media options and viewer and readership data; and, finally, obtain instant feedback on concepts and plans and quickly obtain approvals to move into production. Primitive marketing workstations are already being used today, but the need to match the pace of the design and production cycles will increasingly move the intelligent, decision making tools into the marketing and sales area.

If all things get pretty close to equal in terms of technology—customers and resellers will choose the products they want on the strength of nontechnological attributes: things like loyalty to a company that has been responsive to their needs, that answers the phone when they call.

Varbusiness

from the producing company's view, these things need to be quantified & made part of the product.

Chapter 4

Product Positioning: The Holistic Approach

Pg 46

Thousands of new products enter the marketplace each year. Product positioning gets more and more difficult as the competitive environment intensifies. In industries like the personal computer software industry, establishing a unique position can be very difficult. There are more than 20,000 companies creating and selling software for personal computers in the U.S. alone. They churn out thousands of new products each year. Most of the programs get lost in the crowd, and many never even make it onto retailers' shelves.

It would be a lot easier if product positioning were simply a matter of figuring out what to say about the product, developing a simple slogan, and running lots of ads. But that just won't work anymore.

A new approach is necessary. The idea behind product positioning is simple, but its implementation is difficult and often complex. To gain a strong product position, a company must differentiate its product from all other products on the market. This requires cooperation among the product designers, the manufacturing people, and the marketing organization. Products exist in a constantly changing competitive environment. If the people who design and manufacture the products don't have one foot in the market and one foot in the technology, they won't be able to match the technical capabilities of the company with the opportunities of the market. Message making doesn't position products, actions do. Through its actions, the whole company plays a role in positioning. The product and the market are the yin and the yang of business. The goal is to give the product a unique position in the marketplace. The marketing organization alone can't do that. It takes the whole company.

A company can differentiate its products on the basis of many factors: technology, price, application, quality, service, distribution channels, target audience, specific customers, and alliances. But product positioning extends well beyond the product itself into the perceptions and issues of the marketplace.

How can a company gain a strong position in such an industry? It isn't easy, and it's getting more difficult all the time. But it can be done. When I advise companies on product positioning, I stress eight central concepts.

First, the company needs to understand market trends and dynamics. I often tell my clients they can't position their products by themselves. *The market actually positions products.* But if companies understand the workings of the market, they can influence the way the market positions their products.

Second, the company should focus on intangible positioning factors. Too many companies try to sell their products on price or technical specifications. It is much more effective to establish positions based on soft factors such as quality or technological leadership.

Third, the company must develop the whole product. Products have both tangible and intangible attributes. A computer without software is missing an important tangible piece and is not a whole product. A new product introduced by an unknown company with questionable financial support is missing the intangible elements of the whole product. Computer buyers not only buy the machine but also buy comfort and security. Businesses buy technology leadership along with products because they want the assurance that they have bought the best.

Fourth, the company should target its product at a specific audience. A company shouldn't try to be all things to all people. It should find a niche. Every market, from the automobile to the computer, began as a niche. Perhaps a company should sell its product only to a certain industry, or maybe it should specialize in a particular application of the product. Whatever niche it chooses, the company should then serve the niche better than anyone else in the market.

Fifth, the company must understand success and failure. Most companies never analyze why they succeed or why they fail. I have observed that many high-tech start-up companies fail on their second products. They never seem able to repeat that first success. Companies need to understand the subtleties of success and failure so they can develop a consistent new-product-development and marketing process.

Sixth, it also is important for companies to understand the differences between being market driven and being marketing driven. I learned this lesson many years ago when I worked at a semiconductor company. I watched two product-line managers approach their job in different ways. One spent 80 percent of his time *in* the field calling on customers and prospects. The other sat at his desk and wrote memos, brochures, and promotions which he sent *out* to the field. The first product-line manager was market driven and highly successful. The second was marketing driven and his product line fell to the lowest sales in the company.

Seventh, the value of a brand name is different for complex products than for low-risk consumer goods. This is also true for services. All companies want a distinctive, recognized name. However, even in the world of consumer goods, brand loyalty is dying. Consumers are more willing to experiment with new names on the shelf. That also seems true in the computer industry. "Other" owns the largest share of the personal computer market. A brand must be more than a simple cosmetic icon. It must represent the qualities of the company as well as the product. Having a distinctive brand is more important than ever because the surrounding noise level is so high.

Finally, the company must be willing to experiment. With new types of products, no one can be certain of the best positioning ahead of time. A company should experiment with new products, then pay attention to the market reaction. If users suggest changes, the company must shift course and adjust its strategies.

THE ENVIRONMENT DEFINES THE PRODUCT

From the customer's viewpoint, a product carries with it many meanings. Technology often becomes symbolic. Few people who buy stereo equipment know what the name Dolby actually means about the sound. Few personal computer buyers can distinguish between a 486 or 68040 processor, an OS/2 or MS-DOS operating system, or a CISC or RISC processor. Yet the terms have become symbolic and are used by marketers and salespeople to represent technical differentiation. In technology industries, discussions about such terms take on religious proportions. Products introduced into this environment often become positioned on these religious issues. The discussions and debates within the infrastructure are part of the positioning process. The various alliances that are formed cast light on particular products and help position them as well. MIPS, a start-up in Silicon Valley, set out to challenge Intel and Motorola in the microprocessor-chip business. The microprocessor is the brain in every computer. MIPS developed a new type of advanced processor chip. The multibillion dollar competitors seemed almost too formidable to be challenged in the marketplace. Then, Digital Equipment, the second largest computer company in the world, selected the MIPS processor for inclusion in its next generation of workstations. Overnight MIPS was positioned as a leader. Two years later, Compaq Computer selected MIPS for use in its next generation of personal workstations. MIPS won against the biggest and most powerful of competitors. The marketplace now positions MIPS's products as the next generation leaders. MIPS has yet to run an ad on its new technology but already it is positioned for success.

Channels, distributors, and salespeople also play a role in defining the product. Imagine two brands of wine. Each is made from the same grapes, stored in the same cellar, bottled in the same type of bottle. Identical in every way. It might seem impossible to differentiate one from the other. But now imagine that one of the brands is for sale at discount markets. The other is sold in gourmet food stores and served in fine restaurants. The two are no longer identical. The discount-market brand is perceived as a mediocre wine. The other is seen as a premium wine.

This example shows the power of the market environment. Remember, it is the environment that defines the product. A product can't be viewed in isolation. The elements of the environment—technology trends, market dynamics, competition, social and economic trends—all influence the way customers see the product.

Companies can't just send a positioning message out to the market. They must work with the environment to differentiate and position their products. They must understand who the market influencers are, what the religious issues are, what people are thinking, what their prejudices are, what their likes and dislikes are, what they want to hear. Then companies must position their products to fit in with the attitudes of the marketplace.

Even if two products have identical features and identical prices, customers might perceive them differently. Maybe the company that produces one of the products has a better reputation for quality. Or perhaps it has better technology. Or a more impressive customer list. In any case, the environment makes the seemingly identical products appear quite different from one another.

This is as true for computers as it is for wines. Computers sold through mass retailers have had difficulty being taken seriously by the corporate business markets. In the mass-market retail stores, computers may be perceived as entertainment or game devices rather than as productivity tools.

Timing is also critical, as customer perceptions change with time. A $3,000 home computer would have been perceived as inexpensive in 1977, but as expensive in 1991. As market conditions change, customer perceptions change, and thus product definitions change. Product definition is, to a great extent, in the mind of the beholder.

The central idea of positioning, then, is to use the <u>market environment</u> effectively. Companies should use the environment to make their products seem unique. Marketing expert Theodore Levitt of the Harvard Business School made this point in his article "Marketing Success through the Differentiation of Anything," published in the *Harvard Business Review*, January/February 1980. He explained: "Economic conditions, business strategies, customers' wishes, competitive conditions, and much more can determine what sensibly <u>defines</u> the product. One thing is certain. There is no such thing as a commodity—or, at least, from a competitive point of view, there need not be."

[handwritten margin notes: if they seem unique) they ARE unique. Broader defn of product]

Successful companies always consider the environment when trying to position their products. A case involving Intel provides an example. In the early 1980s Intel noticed that its customers were becoming increasingly concerned about the cost and productivity of software development. To make Intel microprocessors useful to a broader audience, companies needed to develop a vast array of software for them. But software development is a slow, expensive, and labor-intensive process. Companies worried that they wouldn't have the time, money, or manpower to develop all the new software that was needed.

Intel president Andy Grove understood this environment and coined the term "Software Crisis" to describe it. He explained to software companies how Intel products could help ease the crisis. He focused on several new Intel chips in which some software was actually built into the silicon itself, thus reducing the work, and the cost, for software designers. In this way, Intel products were successfully positioned as solutions to the Software Crisis.

As companies see changes in the environment, they must change the positioning of their products. Consider the case of Measurex. In the early 1970s Measurex sold a digital computer to paper manufacturers. Using the Measurex computer, paper manufacturers could produce more paper without increasing the amount of raw material. Customers viewed the product as a productivity-improvement tool, and Measurex reinforced this image through its advertising and sales approach.

But in 1973 the oil embargo caused the environment to shift dramatically. Paper manufacturing is an energy-intensive process, so rising oil prices posed a major threat to industry profitability. In response to this changing environment, Measurex repositioned its product. It ran ads that read: "The Interstate Paper Company will save 100 barrels of oil a day using the

THE ENVIRONMENT DEFINES THE PRODUCT

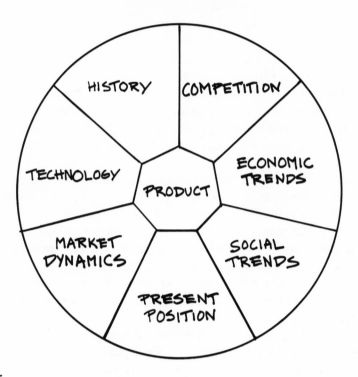

Figure 5.

Measurex. . . ." In essence, Measurex redefined the product as an energy-saving product. Paper manufacturers bought Measurex computers to save energy, and Measurex's sales continued to grow. The product was physically the same as before, but it had a new definition and a new position in the market.

Arm & Hammer is working with Clean Water Action, a national grass-roots environmental group dedicated to protecting America's water resources. The company is giving funds to the group for the rollout of "Home Safe Home: Environmental Shopper's Campaign." The company is also installing "enviro-centers" in supermarkets. Reynolds Metals started a national consumer program of recycling aluminum cans more than twenty years ago. Now it is putting together a consumers' aluminum foil recycling program. Although these companies' products can be seen as having a negative impact on the environment, these companies have undertaken marketing programs to change the way they and their products are perceived.

Other companies should work at redefining their products in the same way. In dynamic industries, the environment usually changes from one year to the next. Companies must constantly monitor the environment to spot changes in customer perceptions and attitudes. Then, like Measurex, they can shift their marketing strategies to fit the new environment.

FOCUS ON THE INTANGIBLES

Companies love to make product comparisons. It is common for a company to boast that its product has the lowest price in the industry, or that it is 25 percent more powerful than any competing product. Indeed, an incredible number of positioning strategies center on price and "specsmanship" (promotion of a product by its superior technical specifications, or specs).

But these approaches to product positioning have their limitations. For one thing, it is difficult to sustain technical leadership. In the technology business, a company is lucky to sustain technical superiority for more than six months. What does it do then? The goal must be to build other qualities into the company's position that can carry it through the ups and downs of the competitive technical horse race. Companies are much better off if they establish positions based on intangible factors, like technology leadership, industry standard, quality, reliability, and service. Unlike price and technical specs, intangibles don't fit neatly onto a product-comparison chart. They can't be adequately measured or articulated by numbers. But intangibles are much more powerful as positioning levers. *intangibles to the customer must be tangible to the company - else you can't control it.*

Why are intangibles so powerful? First, let's take a look at why price and specsmanship are so ineffective as positioning factors. Competing on price has all sorts of problems. Low-price products are often perceived as low-value products, particularly in consumer markets. Consumers assume that cheap in price means cheap in quality. Low price without superior quality can be disastrous. What's more, low-price companies always face the threat that someone else will offer a lower price and steal their position. Further, profit margins often erode along with price, which causes the financial community to say negative things about the company's prospects.

The digital-watch business illustrates this point. The first digital watches, introduced in 1972, sold for about $400. But digital-watch manufacturers kept reducing their prices to underprice the competition. Soon the prices dropped to $99. Then to $49. Then to $20. Sales of digital watches increased. But hardly anybody was able to make a profit, and many companies abandoned the business.

The idea of the learning curve, or experience curve, encourages this type of behavior. A learning-curve strategy involves a two-step logic. First, a company lowers its prices to increase sales volume, thereby gaining market share. Next, the company takes advantage of economies of scale and mass-production experience to cut its manufacturing costs. Prices are lower, but so are costs.

Unfortunately, the logic breaks down when several companies play the game at once. Prices spiral downward more quickly than expected, and profits follow. Makers of semiconductor memories have fought this type of pricing battle several times, to no one's advantage.

Positioning based on specsmanship has similar problems. Companies that position their product as the fastest or the most powerful often run into trouble. Technological leads are usually short-lived. Research labs develop new technologies every day, and new start-ups rush to commercialize them. Products go from being on the leading edge to being obsolete more quickly than ever before. As a result, companies that live by specsmanship often die by specsmanship.

There is another problem: Companies that use specsmanship as a positioning lever often ignore the market environment. They see product positioning as an analytic process of product comparisons. They make huge charts showing that Product A can store fifty more kilobytes than Product B. Or perhaps Product A can perform certain tasks five nanoseconds faster than Product B. These comparisons have some value. But they represent only the beginning of the positioning process, not the end.

In fact, most customers aren't too interested in narrow technical differences between products. Very few people buying personal computers understand the technical differences between one machine and the several hundred others on the market. Moreover, they really don't care. Customers are much more influenced by intangible factors, like technological leadership and product quality, and service and support. It's not easy for a company to position a product in terms of intangible factors. The company must build a certain aura around the product. But if it succeeds, it will attract customers and can charge premium prices.

Sony, for example, has established itself as a leader in consumer electronics. Its color television sets and Walkman products are perceived as the finest available. The Sony position has been established by product miniaturization, design, quality, and innovation. This image carries over to other products. Even though many of its products and innovations have failed over the years, Sony relentlessly continues to pursue its objectives. The consumer remembers only the successes. Only after years of investing in innovative and quality products has the name "Sony" emerged as the leader in consumer electronics. Few people buying a Sony product can differentiate its technology, but Sony's reputation assures them that they don't need to.

The power of intangible positioning became clear to me a few years ago when I was doing a market survey for Intel. As part of the survey I talked to a number of engineers about a certain memory chip. I remember asking one engineer why he selected the Intel chip. This chip was a fairly technical product, and you might have expected the engineer to answer, in technical argot, "The memory had an access time of so many nanoseconds," or "Its power dissipation is only such and such." That didn't happen. Instead, the engineer told me his company buys almost all its chips from Intel, so it was natural to buy the new chip from Intel too. Had he evaluated the new product? Not really. "We just tend to buy from Intel because we have a business relationship there," he explained. "We know where they are going and we trust the company."

I had a similar experience when I visited two small computer manufacturers. Each used Japanese semiconductors in its products. I asked the president of each company why he bought the Japanese chips. The answers were the same: "Quality." I asked whether they had done any comparative testing. The answer was no. I asked whether they did incoming inspections. Again, no. Yet they were convinced the Japanese chips were of higher quality. After all, don't the Japanese have a reputation for high-quality manufacturing?

These incidents are hardly unique. Most buying decisions are made the same way. Product managers spend days, if not weeks, drawing up charts and graphs that compare products on the basis of specifications and price. But buying decisions are rarely based on these objective standards. The important product comparisons come from the minds of those in the marketplace. And in people's minds, it is intangible factors that count.

How can a company gain a position by focusing on intangibles? Intel provides a good example. Intel has succeeded in positioning its products as technology leaders in the semiconductor industry. Intel didn't gain this position by specsmanship. In fact, competitors' chips often have superior specs. Rather, Intel's image as a technology leader is based on its people and its production processes. Top executives Bob Noyce, Gordon Moore, and Andy Grove have a long list of engineering accomplishments, and they often served as spokesmen for the semiconductor industry. Competitors and customers saw Noyce, Moore, and Grove as three of the best technical minds in the industry.

Intel also has convinced customers that its proprietary HMOS processing technology is the best in the industry. The process has an aura developed about it. Even when other companies introduce chips that are faster or denser than Intel's chips, Intel still is viewed as the technology leader in the industry. Specs alone can't dislodge Intel from its intangible positioning.

Intel's position was not easily achieved. The company experienced many failures on its way to its present position. But its failures in DRAMs, consumer electronics, bipolar and analog microprocessors, bubble memories, and CCD technology fortunately did not deter management from investing in new products or from working closely with key customers. Somehow the marketplace never focused on Intel's failures but always on its successes. That has to be the result of its continuous leadership position. When asked why Intel experienced so many failures, Bob Noyce, the charismatic founder, said, "Maybe we should have more failures. That way you'll know we're trying to do new things more often."

THE WHOLE PRODUCT

A few years ago, on a flight from New York City to California, I sat next to the general manager of a division of a large, well-known industrial company. We talked about personal computers. He told me he had mandated that all his managers use IBM computers. I asked why. He said he wanted to be sure that all his divisions' personal computers talked to each other. (This was long before adequate network products were available.) I asked him, "What makes you think IBM PCs talk to each other any more or any less than other computers on the market?" "I *believe* they do," he said. I told him belief was religion not generally considered a part of business.

There are both tangible and intangible attributes to every product. A consumer buys a perception as part of every tangible product. Remember, IBM didn't make a single hardware or software component in the PC. It only added two pieces—the most important: resources and perception. Perception extends from the quality reputation of the product and its manufacturer to the reputation of other established users of the product. But, most important, it encompasses the responsibility for the solution. The more qualitative enhancements ascribed to the product, such as high quality, good service, good support, technology leadership, usage by reputable companies, and so on, the more likely the product will be accepted. Qualitative attributes are derived from experiences with the product and with the servicing organization.

Quality is communicated to the market through an experience. And the communication most valuable in establishing a qualitative position is in the service experience. Customer loyalty begins with an experience. A technology-product concept includes not only the benefits of the product performance but also the integration of that technology or product into the customer's plans and objectives, the future direction of the product, and the economic consequences of the decision to use the product.

Commenting on the bank's computer purchases, former Bank of America president Sam Armacost said, "When we buy, we buy big, so we have to be careful that the vendor has staying power." Perhaps more important than staying power is the completeness of the product solution offered. For established companies with a history of success, the perception of a complete solution will suffice for a certain period of time. But for new companies or companies entering new markets, unless the perception of a product solution becomes a reality almost instantaneously, the resulting word-of-mouth promotion is lost.

FINDING THE RIGHT TARGETS

Product positioning is not based solely on the characteristics of the product, be they tangible or intangible. It is also based on how the product is targeted. Companies can build strong product positions by focusing on specific market segments. Baseball old-timers used to say: "Hit 'em where they ain't." Companies can do the same. They can find segments of the market that other companies have ignored, then hit into the open spot.

Too many companies try to be all things to all people. They want to become $1 billion companies overnight. I've encountered many start-up companies that focus on getting orders rather than on developing markets. They go after and get business in diverse and often unrelated markets, taxing their already limited resources. That is a big mistake. A shotgun approach not only taxes limited resources, but also limits the leverage a company might develop by having a significant piece of business in a specific market. It's better to be a big fish in a little pond than a little fish in a big pond.

In their book, *The Winning Performance—How America's High-Growth Midsize Companies Succeed,* Donald Clifford and Richard Cavanagh studied 6,117 midsize growth companies in America. The companies had, over a five year period including the 1981-82 recession, growth rates four times the growth rate of the comparative quartile of the Fortune 500. Of the companies studied, 74% got their start with an innovative product, service, or way of doing business.

These midsize companies succeeded by targeting niche markets. Their data clearly show that participation in niche markets is more likely to be profitable than is participation in larger markets. The authors point out, "For the midsize company, the niche strategy is born of necessity but cultivated by design. Small and midsize companies simply don't have the resources or staying power to fight head-to-head battles across the board against large, entrenched competition. Instead, they seek out niches that initially are either unknown to larger potential competitors or too small to attract them. Through sheer perseverance and dedication to serving their customers better than anyone else—which to them means knowing the customers and their needs better than anyone else—the midsize companies capture niches and protect them from even the biggest and most formidable companies."

There are two major reasons why companies should target their marketing efforts. The first reason is obvious. A company that targets its products naturally has less competition. As a result, it has a better chance of establishing itself as the leader in the market segment it chooses. The second reason is less obvious but equally important. When a company focuses its efforts on a particular segment, it can do a better job of understanding and meeting the needs of its customers. And that certainly puts the company in a better position to succeed.

Niching is not a new phenomenon. Computers have given us the tools to identify increasingly smaller groups of customers with specific buying profiles. Computers have also allowed us to produce messages specifically directed to narrow groups of users. The entrepreneurial craze of the the '70s and '80s created an enormous array of new products and services. Many companies created opportunities by segmenting markets with narrow product definitions. "Premium ice cream" is an example. As I mentioned earlier, "other" has become a major factor in almost every marketplace.

In the coming decade, I believe we will see more sophisticated niche-marketing strategies. The tools are available to identify customer groupings in almost every conceivable way. Channels to the customers are diversifying as well. For example, the ATM has opened a new channel for reaching banking customers. It has yet to be exploited as much more than a place to get cash but it will be used in the future for all sorts of financial transactions. And it will allow for personalized responses according to the individual location and the needs of the customer.

Even as we speak of globalization, local businesses serving the needs of a small community of consumers are on the rise. Globalization is certainly an important marketing issue for the '90s. However, in the words of Michael

Spindler, chief operating officer and former international marketing manager for Apple Computer, "We must think globally and act locally." Until the giant corporations learn how to act and respond to local, niche markets, they will continue to be nibbled to death.

There is no question that technology will continue to create opportunities for new markets. Investment in new technology is increasing throughout the world. The purpose of expanding that technology is to produce more and different goods and services. Convex, a $200 million supercomputer company located in Texas, began shipping computers only five years ago. These million-dollar machines, more than 600 of them, are being used in chemical and pharmaceutical companies, and by automobile design manufacturers, brokerage houses, electronic companies, and even by an ice cream maker. They are being used to create new processes and products. A great societal energy force is pushing us to discover new things and new ways of doing work. Since these goods and services are being designed to solve problems having narrow definitions and to address specific user needs, companies must understand the mind of the marketplace in order to create new product categories and market niches.

Product positioning is strongest when a company can invent an entirely new market segment as its target. Then, the company can establish a new positioning hierarchy and automatically install itself as the leader of that hierarchy. But in most cases, creating a new segment is not possible. Instead, companies must examine the market environment and decide which existing segments are best suited to their strengths.

Once a company finds the right markets to target, it should maintain that focus as it adds follow-up products. This advice seems logical, but many companies ignore it. Companies often feel an urge to expand into new areas where they have little expertise and no established position. Of course, companies must continue to experiment with new ideas. They can't fall into a rut. But they must remember where their positioning strengths lie and take advantage of them.

Digital Research, Inc. is one company that fell into such a trap. In the late 1970s, the company became a big success by selling system software for personal computers. Its CP/M operating system emerged as an industry standard and the company's profits soared. But Digital Research then expanded into retail application software—low-end application software aimed at consumers. The retail software business is very different from the system-software business, and Digital Research's culture and expertise were poorly suited for the new business. Its expansion efforts flopped, and the company saw profits drop sharply.

TeleVideo is another example. In its early days, TeleVideo established a strong position by selling multi-user computer systems—systems used by more than one person at a time. It sold the computers to other companies, which repackaged and resold the systems. It was in a business-to-business business. The company was a huge success, achieving revenues of $128.3 million in 1984. Then, TeleVideo tried to expand away from its strength. It ignored its established position and tried to sell computers through retail stores. It wanted to build up volume to cut its manufacturing costs. Expenses rose, but volume never increased. Dozens of new computers were competing for retail shelf space, and TeleVideo had no way of distinguishing itself. Retailers were unfamiliar with TeleVideo, and few of them carried the TeleVideo machine. The company never recovered.

> The first principle of management is that the driving force for the development of new products is not technology, not money, but the imagination of people.

David Packard

Chapter 5 | Why Products Succeed; Why They Fail

In his book, *The Fourth Dimension—Toward a Geometry of Higher Reality,* Rudy Rucker describes a book published in 1884 and authored by Edwin Abbott. This book tells of a land of one-dimensional characters and a square that takes a trip into a higher dimension. The place is called Flatland. Abbott's book is a satire on nineteenth century society, but it presents a thought-provoking story that may well be analogous to the state of the technology industry today.

"An initial question about Flatland," says Rucker, "is the problem of how these lines and polygons can see anything at all. If you were to put a number of cardboard shapes on a table top and then lower your eye to the plane of the table, you would really see just a bunch of line segments. How can the Flatlanders tell a line from a square? How can they build up the idea of a two-dimensional world from their one-dimensional retinal images?" In the past few years we have seen the creation of a "technology Flatland" with thousands of one-dimensional products exhibiting few, if any, intrinsic or perceptible differences. Little wonder that few survive.

Currently, there are more than 200 IBM PC clones, 240 or so suppliers of personal computers, more than sixty 5¼"-disk-drive companies, more than 150 word processing packages, and more than 100 spreadsheet and database packages. It has been estimated that there are as many as 16,000 software producers. In the past five years 5000 software companies produced 27,000 different software packages.

Perhaps as much as $5 to $10 billion is spent on development of new technology products that fail. This estimate may be conservative when marketing, sales, promotion, and time lost expenses are included.

This problem isn't unique to technology companies. *Gorman's New Product News* reported that 13,244 new products were introduced in the United States in 1990. It is estimated that between 70 and 80 percent of those will fail.

No one factor causes products to fail, but certainly one overriding factor is that these products are not market driven. Differentiation is only in the minds of the developer. In this "product Flatland" only its own inhabitants can perceive any differences in shapes. To the outside world the products remain one-dimensional.

In *The Fourth Dimension,* Rucker tries to get us to think beyond the limits of our self-created boundaries. He asks us to "think of 'dimension' as being any possible type of variation, category or distinction." Rucker suggests that there is no reason to limit the "dimensions" of the world to space and time. "Part and parcel of every object you see is what the object reminds you of, how you feel about it, what you know about its past, and so on. If we can make an honest effort to describe the world as we actually live it," says Rucker, "then the world grows endlessly more complicated than any simple 3-D picture. There is a feeling that the more we delve into the nature of reality, the more we find. Far from being limited, the world is inexhaustibly rich."

I think this passage is great because it's so dynamic and so descriptive of the world of technology. Developers of new technology products need to seek a new dimension, a dimension that lets potential customers see, feel, enjoy, and be excited by the new application of technology.

We will never reach that dimension using promotional techniques. It will emerge and flourish only in the presence of imaginative new products whose technology is integrated into customers' everyday lives.

Ideas and opportunities have always been plentiful and there is no reason to believe that the imagination of mankind has been exhausted. We must understand that marketing is a learning experience. We have to apply quickly the lessons we have learned. Education simply is not valuable unless it is applied. Product success is the result of an effective dialogue with the marketplace. Failure comes when a breakdown occurs in that communication process.

I first got interested in the subject of product success and failure when I noticed that so many "second products" of entrepreneurial companies fail or don't live up to market expectations. To better understand product failure I analyzed a large number of products in the technology marketplace, but I believe my findings apply to many other industries and businesses as well. I used the history of the Intel 8086 microprocessor, the Apple II computer, the IBM PC, the DEC VAX minicomputer, Lotus 1-2-3 spreadsheet software, and other products for the basis of my observations.

From the time it's only a glimmer of a concept, a product's success is in its genes. Products don't suddenly appear in our midst and overwhelm us. Most successful technology-based new products have certain characteristics:

innovative application of technology.

1. They appeal to new markets or an expansion of existing markets so that thousands of new and potential customers for the products are created.

2. They are not inventions, in the true sense, but assemblages of inter-related technologies—their creators take advantage of existing technology, achieving leverage by combining what they find in the world around them into a unique package. Product innovators are students of the market and the technology. Successful products are created by the addition of a new dimension to existing products and markets.

3. They are dependent upon other newly developed technologies and the market infrastructure for their success. The Apple II required the use of the first miniature disk drives and the 6502 microprocessor, which both enhanced and sped the software development. Lotus 1-2-3 needed the IBM 512K PC as a platform. Whole new markets are born out of this interdependency because companies are forced to work like partners, pooling their technologies and talent. In his book _Inside the Black Box: Technology and Economics_, Nathan Rosenberg states that "the growing productivity of industrial economies is the complex outcome of large numbers of interlocking, mutually rein-forcing technologies, the individual components of which are very limited economic consequence by themselves....

   "...Many of the numerous instances of entrepreneurial failure can be attributed to the fact that a would-be entrepreneur failed to consider the relevant conditions of interdependence between the component with which he happened to be preoccupied and the rest of the larger system."

4. They are timed right. Ideas for most successful new products are not particularly original. The IBM PC was modeled on the Apple II, the Apple II on a variety of hobby computers, Lotus 1-2-3 on VisiCalc, Macintosh on the Xerox Star. Businesses that focus on developing products and markets before anyone else usually win. This is why having one eye on the market and one eye on advances in technology is such an advantage. Successful innovators see what's going on and do something about it.

5. They are adapted to market requirements. Technology products must be molded to the markets. They must be adapted, enhanced, and

1. suggest the "right time" is simply the time when a "complete" product is available.

incrementally improved. Success comes to those who keep improving and adapting their products. The Japanese automakers are a shining example. The first Honda that came into the United States in 1970 was a far cry from the new Acura. And consumer products from Sony and Panasonic just seem to get more and more refined, with added features, better quality, and greater reliability year after year.

6. They are developed by small teams. A sense of purpose and mission can be better generated in a small community.

7. Key customers often play a central role in their design and incremental improvement. As we move into the future of custom-designed products, the customer will be integrated in some way in the design process.

8. The first users determine their success or failure. Early users become advocates and help sell new prospects. If the first users are involved in early definition and development stages, they usually take a large share of credit for the product and want to insure its success.

9. Revolutionary new products generate new language. Icon, mouse, windows, Apple, menu, and Macintosh all are words with noncomputer origins that have become part of a new technology's lexicon.

10. They are used in seminars, workshops, and demonstrations, and are discussed in newsletters and at conferences, to educate and develop the market.

Other things such as good luck, good management and competitor error also help make these products successful, but continued investment in R&D is the best way to insure product success in the marketplace.

On the other hand, failure is on the flip side of the coin from success. More often than not, failure is the result of poor management and inability to efficiently implement product development. I have observed a number of company and product failures over the years and, in poring through my notes, have noticed that they seem to have several things in common:

Your complacent - comfortable w/what you know

1. The product doesn't create or expand the market. Generally the product has undergone an incremental technical improvement, with little or no cost/performance benefit. It ends up competing with existing successful products.

take the easy way out - but you know you should build a 2nd generation product.

2. Indecision is reflected in the product design and offerings. Management can't decide whether to incrementally improve the present product generation, displace it, link the two products as a family, go up or down the spectrum of performance and cost, or build something radically new.

3. The people who developed the first product have moved on or are no longer in a position to judge the technology and market posture.

4. Due to the success of the initial product, arrogance sets in. Success blinds management to the possibility of failure. They convince themselves that the market will embrace whatever they create. *Not as much external input. Roger Smith, GM, "There isn't anything*

5. The idea for the first product occurred in an environment that is no *I don't* longer present. The decision-making process is different. Often an *know* entrepreneur developed the original idea while working at a larger *about* company where technology and ideas abound but choice is limited. *building* The PCjr from IBM, the Apple II, or Symphony from Lotus are high- *cars."* profile examples of second-product failures. Learning to build a successful new product development process is a cultural issue.

6. Democracy decides the product. The new team feels that everyone involved must have a say in the product direction.

7. The company loses track of the market guiding influence—in which customers interact with the product-development team.

8. Promotional marketing techniques replace relationship marketing and market development.

New-product success depends more than ever upon savvy marketing because companies must have reliable market knowledge in order to appropriately adapt their technology. Marketing should be the guiding influence for companies so that they can adapt, change, and quickly respond to the needs of the customers and the market.

I once saw George Burns interviewed on a television talk show. Among his many talents he had to be a good marketeer. The interviewer asked George, "What made you and Gracie so successful?" George replied, "Our audiences did it." The interviewer acknowledged George's modesty and insisted that it was the genius of talent that made them successful. "No," said George, "we had a number of routines in the early days. We experimented a lot. Gracie, for example, had a routine in which she was very sarcastic and

(margin handwriting: 2nd Product Syndrome; Pg 56, 72; principal architect)

Figure 6.   **BY MAL**

one in which she played the scatterbrain. The audience did not respond to the sarcastic routine but they loved the scatterbrain. So we dropped the sarcasm and built on the other."

Talent is a given in any successful business. But the *willingness* as well as the *ability* to read your audience is what marketing is all about. More than anything else, the inability to accept change and touch the customer in a meaningful way inhibits the growth of the technology business.

ORIGINAL CONCEPTS, INCREMENTAL IMPROVEMENTS

The problems facing the technology industry today are not those of technology. They are problems of limited dimension.

The dimension we have to find is one that will bring new users of the technology into the market. That dimension isn't created by our assuming that the light at the end of the tunnel is a pot of gold waiting to be discovered.

It is a dimension that must be built: built on imagination, education, development, and adaptation of the technology. It brings the user into the design process. To succeed in the future, marketers will have to learn new and better ways of bringing value to the customer. Markets are not infinite resources to be tapped. Markets grow, change, and develop in response to innovative products and innovative ways of doing business.

There are two general kinds of innovation: an original concept and an incremental improvement. Both can be market creating. The Polaroid camera, for example, created a new market for cameras. An original concept, it established not only a new class of products but also a new class of users. An incremental improvement can create a new market if the improvement is such that the innovation brings new users into the fold. Lotus 1-2-3 was a significant improvement over VisiCalc and brought nonfinancial types into the spreadsheet-user market. When the automatic transmission system was incorporated into the automobile significant numbers of people began driving.   *ie. the product*

Too many innovations today are nothing more than feature improvements. They don't extend the technology to reach new audiences. They don't create new markets. Innovations that don't define for themselves a new user group must then go head to head with the established competitors in a market-share battle. Small companies whose managements try to build their businesses on such innovations will find themselves doomed to fighting market-share battles with resource-intensive competitors.

Technology products are not born perfect. They need the proper environment and time in which to grow, change, and adapt. The success of any technology product depends upon the ability to change and adapt to various market requirements. Perpetual adaptation makes a product successful. All successful computers go through a series of incremental improvements and market adaptations throughout their life cycle.

Architecture designed for change.

The Apple II's success, for example, is directly related to flexibility and the capability of third-party software and hardware vendors to adapt the product to many different markets. Also incremental improvements were made to enhance its performance and reduce manufacturing costs. The IBM PC adapted, changed, and likewise found success. All successful minis and mainframes are adaptable to markets by third-party software houses, vertical-market specialists, system's integrators, and a wide range of peripheral-product and software vendors. This adaptability is what makes a technology or product market driven. Marketing must focus on and address this adaptation process, not just the delivery of products.

MARKET DRIVEN OR MARKETING DRIVEN?

The term *market driven* has become popular these days. But I'm afraid most people don't know what it means. Some think it means surveying the market, and then making products in response to what customers want. The term *market driven* carries with it some implication that the marketplace tells the company what to make and how to act. Science and technology create many new things with no initial perceived value. When the first airplane, automobile, and personal computer came to market few people saw themselves as potential customers. However, once the products are in the marketplace, a "dialogue" takes place between producer and consumer. The market pushes back, it lets its needs be known; the technology changes and adapts.

The electric starter and automatic transmission for the automobile made driving easier and so expanded the market for cars. Packaged software made personal computers easier to use and thus the market expanded. In this sense, the dialogue directs the *refinement* of the product. Product positioning is not just a matter of inventing and offering a unique product. It is much more a matter of a company adapting its technology rapidly in response to the dialogue with the customer.

The product-positioning process is the packaging of the "total solution" for a given group of customers or marketplace. Technology innovates, but the market refines. The ability of companies to adapt to the changes in technology and in their products in response to customers' needs is what is meant by the term *market driven.*

Market-driven products appear more like services than products. The computer, PBX, microprocessor, computer network, laser, modem, and other technology-based, complex products are at the same time specific and general. The generic products are customized through changes made by software, support, and/or service organizations. For every one engineer designing a computer, for example, there are ten people providing the adaptation, service, and support necessary to fulfill the customer's requirements. Such products require extensive application knowledge and "personalization." Market-driven products rely on an understanding of the customers' needs, the competitive environment, the industry's religious issues, word-of-mouth, infrastructure acceptance, adaptability, cost effectiveness of the solution, and the credibility and reputation of the supplier for their success. *Experimentation.*

Marketing-driven products, on the other hand, are less adaptable and are based on a narrow product definition. That is, the products have specific

---

78

and limited definitions. Further, there is little risk associated with marketing-driven products. Products such as toothpaste, soft drinks, calculators, some electronic components, or any traditional commodity are low risk to the buyer. Changing brands of soft drink requires risking less than a dollar. Changing or acquiring a microprocessor, data-base software package, PBX, or computer system can cost plenty, not only in terms of money but also in terms of time, future changes, and ultimately, competitiveness.

BRAND

Unfortunately, most of our ideas on brand are derived from consumer marketing strategies. I say it's unfortunate because there is a big difference between the way computer marketing and consumer marketing use brands. I define *brand* as an icon with virtual memory. We see a particular brand and it conveys unseen information. The information is in the mind of the consumer. How did it get there? Through some particular experience. What is its purpose? Awareness.

Consumer marketing is based on the principle that awareness leads to trial purchase. Given the right price and distribution, and backed by a fair amount of promotional effort, it generally does, because the risk of purchase is low. Trying a new brand of cereal, switching from one kind of cola to another, or buying a new brand of jeans is virtually risk free. If I don't like the product or experience, I have lost very little. Recent studies show that consumers are experimenting and changing brands more frequently than ever.

On the other hand, purchasing a personal computer, data-base engine or software platform, microprocessor, laser system, MRP system, LAN, or almost any technology product is high risk to the buyer. Buying a gate array or custom chip, a communication network, or almost any technology product means buying the company behind the product as well. When I buy a technology product I am concerned about the economic stability of the supplier, the installation and development time, the education and training investment required, the cost of ownership, the plans for the product's next generation, the openness of the system and its interoperability, as well as who else is committing to the product, how many third parties are supporting it, and so much more. Because of our concerns, we consumers often are more aware of failures in the technology business than we are of successes. We instantly recognize the names of failed products and financially troubled companies.

Conversely, when we buy a consumer product we never ask about the financial status of the producer. I doubt most consumers know whether

General Mills or Procter & Gamble are making money. In fact, Americans continued to buy cars from Chrysler when it was on the edge of bankruptcy. No one would make such large purchases from a troubled technology company. Technology start-ups as well as businesses that go through financial difficulties have a lot of trouble convincing customers and prospects to have confidence in the future of the company and its products. And consumers don't necessarily lack confidence because they believe the company might go out of business. When technology companies get into financial trouble, they cut back on product support and R&D, and eliminate product lines, sending a message of deep concern into the marketplace. Consumers don't think about the financial stability of a packaged goods company while shopping in a grocery store.

The competitive strategy of Coke is "Beat Pepsi." It's not real complex. Coke doesn't have to worry about a new competitive challenger every six months, a new generation of products every two years, pushing the state-of-the-art performance in every new product generation, reducing costs continually, retraining and educating existing sales channels and customers, convincing third parties to jump on the product's bandwagon, and so on.

Brand plays a different role in marketing high-risk products. With low-risk products an established brand is accepted at face value and leads directly to trial and purchase. With high-risk technical products brand is seen in a multidimensional way. A complex, high-risk product brand carries with it the history and performance of the company, the technical leadership, the "camp" (the following created by the technology or product), the users, the quality and support of the product and the company, and many other things. For low risk products brand choice can be based more on cosmetic factors. The package color is important, and so is the latest sports hero or entertainer chosen to endorse the product. Imagine shopping in a grocery store and having to select food brands based on the financial performance of the manufacturer or on compatibility of related products. Brand in regard to a technology product is important in terms of expected performance, quality, service, and support. Consumer marketing is no less difficult than technology marketing. But it is different.

EXPERIMENTING AND CHANGING

Product positioning is not a one-time event. It is an ongoing process. A company president once said to me: "How can we ever position ourselves in a marketplace that is changing every three months?" He had a good point.

Dynamic positioning is a difficult process. The only way to survive in dynamic marketplaces is to keep the positioning process flexible. Companies must be willing to experiment and learn and change. There is no right and no wrong. In fact, as we have seen, the path to success is often filled with failures. Kamran Elahian, founder of two successful high-tech companies, considers "success management of failure. I haven't worked with a smooth operation yet. The objective is to make progress in spite of failures."

Marketing people like to think they know their market. They do analyses of the market, then develop detailed marketing plans as though the outcome is decided deterministically. But in fast-changing industries companies often are breaking new ground. No one can really know the market. The market doesn't even exist yet.

In these industries, almost all new products are experiments. Few leading-edge products are perfectly in tune with the market when they first come out. Instead, they are modified and altered once they meet the market. There's a lot of give and take.

In some ways, the positioning process is the mirror image of the marketing style of traditional consumer businesses. In traditional businesses companies survey people to find out what the people want, then create a product to fill the need. In technology-based industries the product usually comes first. Companies invent things and develop things. Then they work with the market to see how these products should be used.

The process might seem backward, but in some cases it's the only way it can be done. With ground-breaking products, customers can't know what they want until they've seen the product. But after trying the product, they can suggest modifications so the new product or technology fits their needs. Rapid adaptation to early customer needs is critical to success. Osborne Computers took advantage of slow-moving IBM and created the first portable personal computer. But the company vanished when it, in turn, could not respond fast enough with the next generation of its own product.

When the first personal computers came to the market, people didn't know how they might use the new machines. A market research study would have shown very little demand. But a few pioneering companies put personal computers on the market, and people came up with suggestions making the machines better and putting them to new uses.

The same thing happens with many new semiconductor chips. Semiconductor companies rarely go to customers and ask: "What do you need?" Rather, they take spec sheets to certain key customers and say: "Here's what we can do. How should we modify our design to suit your needs?"

Successful semiconductor devices often go through ten or twenty revisions during the life of the product. The experimentation never stops.

Beta sites, the locations where a company first tests its product, can be critical in this process. By working with beta-site customers, companies can begin making modifications to the product before taking it to the market. Oftentimes beta-site customers will hate the product initially. But as their suggestions are implemented they fall in love with the product. It has become just what they need. When the product is finally introduced to the marketplace, it is much more likely to make a good first impression. And that's important, as you never get a second chance to make a first impression.

Experimenting, however, can't end with beta sites. Companies must continue to modify their products and strategies after the product is already on the market. The environment keeps changing, and companies must adapt.

Selling into a changing environment can be tough on the nerves. I once did some work for a large company that was launching a product into a new area. The company's managers didn't know how to price the product, and kept saying how insecure they felt with it. But that's only natural. What they were doing had never been done before. The president of a large Japanese electronics company once asked me how they might determine the market for a new credit card–sized translator. They needed to know because they had to plan the production quantities well in advance. Several market research companies came up with different answers. I suggested they make their best judgment and then "make the market happen." You can target a small segment and build a model. But that comes with the risk that if the market explodes, the competition can move into a market you have created. You can't be secure unless you have a proven model. In new product areas there are no models. You have to experiment. The people who usually win are the ones who have the guts to move forward.

All products must be seen as experiments. Many products and services go through the cycle of failure, change, failure, change. Failure is not necessarily a problem. The critical issue is how quickly a company can respond to consumer reaction. Managers first must monitor how the market reacts to their product. Then they must modify the product or service before competitors come up with their own solutions to the market's needs. The stakes are great. Whichever company modifies its product quickest and most effectively will win the product-positioning battle.

Architecture designed for change.

| **Our audience made us successful.** → *Pg 75/76*

George Burns

Chapter 6

Market Positioning: Developing Relationships

CREDIBILITY

If you looked just at the numbers, you would think the Apple III computer got off to a pretty good start. The machine, aimed at small businesses, was introduced in 1980. By the end of 1983 the annual sales of Apple III computers topped $100 million. That's not a bad year for a product in a young industry. The Apple III outsold many other computers on the market.

But the market saw the Apple III as a loser compared to other products in the market. Many people expected the Apple III to be as successful as the company's first major product, the Apple II. The new computer never lived up to those expectations. Because of some early manufacturing glitches and a lack of software, the machine got off to a slow start. It acquired a bad reputation, and it never was able to get rid of its negative image. Further, Apple did not use the infrastructure to leverage marketing. Apple chose to develop much of its own applications software for the Apple III rather than use the hundreds of independent software vendors in the infrastructure to create markets and sales for the product. Many users were very happy with the machine. The manufacturing problems were solved and the software was gradually appearing. But the Apple III never acquired a strong product or market position.

In market positioning, the second phase of the positioning process, the marketplace responds to the new product. The company must use the leverage of the market to create the product's position. Working with selected customers, using customer advisory groups, making use of beta sites, and working with the third parties in the infrastructure before the product is formally introduced establishes advocates early. The company finds out during this early phase whether its product is positioned well. Winning a quick endorsement from the market is critical to success. You can't wait until the product is in production and shipping to find out if you have infrastructure support for it. Once a product wins over the infrastructure, it picks up momentum in the broader marketplace. Success builds on itself. The product develops a positive image, and the customers follow. On the other hand, once the market sticks a "loser" label on a product, it has a tough time overcoming its unfavorable reputation.

Clearly, companies have little direct control over this stage of the positioning process. Market positioning is determined largely by the perceptions of those in the infrastructure marketplace. This is where the product position is established.

But it is possible to influence the market-positioning process. By understanding the workings of the market, companies can influence the market's perception of their products. They can create a stronger image for their products. They can take steps to make themselves and their products more credible.

Credibility is the key to the whole market-positioning process. With so many new products and new technologies on the market, customers are intimidated by the decision-making process. Many customers don't even understand the technologies used in new products. Technology-based products are links in a chain: they are attractive because they are linked to the future. But when people are buying a piece of the future, they need to be reassured. They want to buy from a supplier with credibility.

Quite simply, customers are often confused by the deluge of claims and counterclaims made by various suppliers. To make matters worse, some large companies play on customer insecurities in an effort to scare customers away from smaller competitors. IBM turned this fear-raising game into a central element of its marketing strategy. The strategy has become known as FUD: Fear, Uncertainty, and Doubt. IBM salespeople build on the fears that already exist in the marketplace. They portray IBM as the only safe haven in an unpredictable, stormy environment. Their argument can be powerful: Why risk buying from a smaller company? No one has ever lost his or her job for choosing IBM as a vendor. FUD can be a powerful marketing tool if the products, systems, or service at stake carry high purchaser risk.

To establish market positioning, companies in fast-changing industries must find ways to ease customer fears and offset the FUD strategies of corporate giants. They must offset Fear with Comfort, Uncertainty with Stability, Doubt with Confidence. They must build images of credibility, leadership, and quality. They must supply the customer with a "security blanket" as well as a top-notch product.

How can companies build credibility with customers? Advertising, of course, can play a role. But advertising doesn't get to the heart of the matter. Advertising can only reinforce products' positions, it can't create them. Increasingly, people are skeptical of what they read or see in advertisements. I often tell clients that advertising has a built-in "discount factor." People are deluged with promotional information, and they are beginning to distrust it. People are more likely to make decisions based on what they hear directly from other people, including friends, experts, or even salespeople. These days, more decisions are made at the sales counter than in the living room armchair.

Advertising, therefore, should be one of the last parts of a marketing strategy, not the first. Sun Microsystems became the leading supplier of computer workstations even though their competitors outspent them ten to one in advertising. Companies in technology-based businesses need other ways to build credibility. They must seem secure and trustworthy to customers who are intimidated by technology. And they must build a solid foundation—one that will survive the inevitable changes in the market environment. Most technical companies forget what made them successful in the first place—relationships. ①

② Of course, companies must start with strong product positioning. Then, they can build credibility—and market positioning—in several ways. The three most important ways are by inference, by reference, and by evidence.

Association

Inference. If a start-up is funded by reputable financial backers, or if the start-up has a relationship with a respected large company, people infer that the start-up must be a credible competitor. Retailers, distributors, and customers begin to take the start-up seriously. Compaq and Lotus gained instant credibility because Ben Rosen, a well-known venture capitalist, was a lead investor in each company. A deal with IBM also brings instant credibility. Few people had heard of MIPS Computer before Digital Equipment chose the small California company to supply its RISC processor for the DEC workstations. Now everyone in the field sees MIPS as a technological leader. Similarly, a deal with AT&T and Sun Microsystems, and a deal with Eli Lilly, brought credibility to Genentech.

Reference. When people shop for complex or expensive products they often rely on personal references. They'll look for a friend or colleague who has purchased the product, and ask whether the product is satisfactory. Anyone who interacts with the product or the company is in a position to act as a reference in the future. Analysts, retailers, journalists, and customers all talk to one another and spread word about a product. If one person has a good experience with the product, that person will tell others about it, and they in turn will tell still others. Credibility builds and builds. But the process works in reverse as well. A rule of thumb: A customer who has a good experience with a product will tell three other people. A customer who has a bad experience will tell ten other people.

Evidence.   Success in the market reinforces itself. People in the industry look for tangible evidence that a company is doing well: rising market share, rising profits, more retailers carrying its products, new ventures, major alliances. Each piece of evidence adds to the company's credibility and image. As Jim Morgan of Applied Materials said, "Image is just a collection of the things we do." Without the evidence, positioning is hollow.

There are many things you can do to put these three credibility-builders to work for your company. For example, you can develop favorable relationships with key people in the industry infrastructure. You must begin by carefully selecting the first users of the product, followed by key third-party support organizations, value-added resellers, and distributors. All of these entities, as well as industry consultants, analysts, and journalists, play a powerful role within your industry's infrastructure. They serve as important references and spread word about the company. By winning over the infrastructure your company will be 90 percent of the way toward winning the market-positioning battle.

Customers can help a company gain credibility by both the reference and the inference methods. By choosing its customers carefully, rather than simply selling to all takers, a company can control its image. If a small company sells products to a respected corporate giant, such as American Express, AT&T, or General Motors, the small company probably will seem reputable and solid. If it sells to fast-growing start-ups, it could build an image as an innovative supplier.

Companies can also gain credibility by forming strategic alliances with other companies. Genetic Systems, a biotechnology company based in Seattle, initially seemed similar to many other biotechnology start-ups— strong in science skills, but weak in business skills. Analysts wondered whether its products could ever succeed in the market. But when Genetic Systems teamed up with Syntex, a large and successful pharmaceutical company, it was suddenly perceived as a legitimate competitor in the medical-diagnostics industry.

Building credibility is a slow and difficult process, but it can be done and it is critical to market success. Let's examine ways in which companies can build credibility, and, in turn, establish market positioning for their products. The strategy can be broken down into four key elements:

- Using word of mouth
- Developing the infrastructure
- Forming strategic relationships
- Selling to the right customers

USING WORD OF MOUTH

When the first Mac factory opened, Steve Jobs invited the presidents of all Apple's suppliers to a dinner and a tour at the factory. I sat with presidents of a half dozen major manufacturing companies. Over dinner, they began discussing the telephone systems their companies used. Two or three of the presidents were looking to buy new systems at the time. During the conversation, one of the presidents made a remark about a certain company that supplies telephone systems, a company I'll call Company X. "Company X is putting me out of business," he complained. "It seems like the system is always down."

Well, Company X happens to be a very successful supplier of telephone systems, and that company has hundreds of satisfied customers. But that one comment probably made a decisive impression on the other executives at the dinner. Company X might have the best systems on the market. Its systems might handle more lines than any other competitive product. They might integrate voice and data better than any other system. But the presidents at that dinner now are unlikely to buy a system from Company X because that one offhand comment made them all feel a bit insecure about Company X.

Thus is the power of word of mouth. Forget about market surveys and analyst reports. Word of mouth is probably the most powerful form of communication in the business world. It can either hurt a company's reputation, as in the example above, or give it a boost in the market. Word-of-mouth messages stand out in a person's mind. Memorandums might contain all the correct information, but face-to-face communication is much more likely to gain commitment, support, and understanding—it is more likely to be believed and remembered. Quite simply, we find messages more believable and compelling when we hear them directly from other people, particularly people we know and respect. We use word-of-mouth communication to help us make all sorts of decisions. We rely on word-of-mouth information when determining what products to buy, what companies to trust, what written reports to read, what corporate leaders to believe.

Information and communication are not the same. Information is cold, objective data. Communication is experiential and qualitative. It is personal. A computer terminal can convey information, but only people can communicate. For communication to be effective, the sender and the receiver must be in sync—on the same wavelength. When people meet face to face and use word of mouth, they are more likely to communicate effectively. Each person can read, analyze, and interpret the attitude of the other person. The same information could be transmitted in a telegram or a written report or an advertisement. But those forms don't utilize the full potential of human communication. Their use as communications tools is limited. Word of mouth turns raw information into effective communication.

Word-of-mouth communication can take on many different forms. Industry participants form "old-boy networks" to keep each other informed about new developments. One recent market research report showed that such networks play a key role in the telecommunications industry. Gaining access to the networks is critical to success.

Customers use word of mouth too. People are confused about products developed by technology-based industries and they want personal information and advice. Hardly a day passes without someone asking me that familiar question: "What personal computer should I buy?" Hardly any computer of any size is sold these days without some word-of-mouth reference preceding the transaction. This kind of information gathering is not unlike the way we choose a family physician or attorney. When the risk is high, we seek advice.

In a KPMG Peat Marwick's *World* magazine article, Jim Easton, head of Jas. D. Easton, the sports equipment company, said "In an industry where names are 90 percent of the game, credibility is everything." The article went on to point out that "even though the company does very little advertising, it benefits greatly from word of mouth. Last year, for example, the company paid only four National Hockey League (NHL) players to use its sticks. By spring, 36 NHL players had Easton equipment on the ice."

Word of mouth is so obvious a communications medium that most people don't take time to analyze or understand its structure. To many people, it's like the weather. Sure, it's important. But they think they can't do much about it. We never see a word-of-mouth communications section in marketing plans. As management expert Peter Drucker once noted, "More business decisions occur over lunch and dinner than at any other time, yet no MBA courses are given on the subject."

The 5th Wave   By Rich Tennant

"I GUESS THERE'S A 'USERS GROUP' FOR JUST ABOUT EVERYONE THESE DAYS."

Figure 7.

Of course, much of the word-of-mouth communication about a company and its products *is* within the company's control. A company can take steps to put word of mouth to its advantage. It can even organize a "word-of-mouth campaign." Such an effort can be very powerful, because word of mouth is fundamentally different, in three major ways, from other forms of communication.

First, it is an experienced process. The message is always carried by a real person. How knowledgeable are the people who represent the company or the people who interpret technology and the product for the marketplace? The product is judged on this factor. The product is also judged by how knowledgeable, articulate, helpful, and diligent representatives of the company are. This personal representation must extend from all who directly or indirectly touch the customer. I'm not talking just about salespeople. The process involves developing enthusiasts of employees, distributors, resellers, university researchers and professors, government officials, financial analysts, and journalists. In complex and confusing market situations, the customer relies heavily on advice and references. The words of a trusted, knowledgeable representative carry more weight than words transmitted by any other form of media. The product message could be sent over telephone lines, but it would not carry the same weight. Businessmen recognize this fact, and so travel thousands of miles just to meet face to face with salespeople and customers. A face-to-face meeting can have greater impact on sales than an avalanche of advertisements, press releases, memos, and brochures.

Second, the message is tuned to the individual listener. The word-of-mouth message can be changed, simplified, altered, embellished, and verified for each person. The message delivered to the director of marketing can be different from the message delivered to the director of engineering. The message can be altered in order to serve the setting as well. A message delivered one way in the company cafeteria would be delivered another way in a presentation to the board of directors.

Third, feedback is instantaneous. The listener, when in agreement with the speaker, will nod or show some other sign of concurrence. In disagreement, the listener will scowl or suggest alternative arguments. Weak points can be fortified and irrelevant ones can be eliminated. If the listener does not understand, further explanation can be provided immediately.

So how can a company harness the power of word of mouth? The company must decide who the message carriers will be. Advocates can be enlisted by selecting recognized authorities or key customers to serve on advisory boards or by letting these people participate in early product testing. Their participation must be desired for the value of their advice and contribution rather than viewed as a promotional activity. If the people who stand between the product and the customer are given a sense of participation, they will be more likely to become enthusiastic advocates.

The nature of word-of-mouth communications makes it impossible to spread the message very widely. Luckily, there is no need to. Word of mouth is governed by the 90/10 rule: 90 percent of the world is influenced by the other 10 percent. So if a company can reach the critical 10 percent, it will indirectly influence all the others. As the critical 10 percent pass the word on to others the word-of-mouth message will grow like a snowball rolling downhill.

A word-of-mouth campaign should be based on targeted communication. Word of mouth is not an efficient means for distributing information widely. Few businesses hold to the notion of being all things to all people in a mass market. Communications should be directed toward specific audiences or market segments. Each segmented market is a system of interrelated organizations. A word-of-mouth campaign aims to develop or change the attitudes and opinions of the people in these target groups, so understanding the market system of each segment is very important. Without this understanding, word of mouth is not likely to be effective.

The targets for a word-of-mouth campaign fall into several categories. The possible targets include:

Customers. Companies can use word of mouth to reach customers at users-group meetings, trade shows, technical conferences, training programs, and industry association meetings. Beta-site programs and early customers become especially important. They become the early adapters and advocates that will be referenced by your salespeople and then looked to by potential customers for reference. Every customer is important, but your first customer can make your product a booming success or a dismal failure.

The selling chain. The selling network includes sales representatives, distributors, resellers, and other third parties who help to bring the best product possible to the customer. Theodore Levitt many years ago wrote an article published in the *Harvard Business Review,* March/April 1966, entitled "Branding On Trial." Levitt advised that the way for the small company to beat the large, well-established company with a large advertising budget is "to get to the consumer last." Training, developing, and enlisting those people who meet the customer face to face pays huge dividends. Having few salespeople and a meager promotion budget can cause a company hardship. Small and medium-sized companies can greatly improve their chances of success by investing in the infrastructure of their industry.

Industry watchers. Rapid-growth industries are filled with consultants, interpreters, futurists, and soothsayers who sort out and publish information and expound in conferences about the industries. These industry watchers gain most of their information through word of mouth—they visit companies, attend analysts' meetings, and talk to others in the industry. Most of these people try to be objective and unbiased, but they are only human. They were invaluable in supporting Apple Computer in its early days. As Apple grew and sometimes faltered, the industry watchers called Apple executives to give advice and help. They continued to write about Apple's products and strategic direction with honest concern and interest. They were Apple fans who didn't want to see that company fail.

The financial community. Who backs a company is often more important than how much money is behind it. The community of venture capitalists and private investors is a small, close-knit group. A company's initial backers can use word of mouth to spread the company's message to other venture capitalists, and later to investment bankers, analysts, and brokers. Financial analysts are also industry-watchers. Every company must either raise or borrow money or deal with Wall Street throughout its corporate life. For those reasons, the financial community is an integral part of every industry infrastructure.

The press. More than 90 percent of the major news stories in the business and technical press come from direct conversations. All journalists have networks of sources they use when seeking background information, opinions, and verifications. Journalists love to talk to customers—the users of the product. Because most markets are narrowly defined today, journalists can easily dig into the infrastructure of any particular marketplace and talk to the knowledgeable as well as the opinionated but less informed. Although a great deal of money and time is spent on press releases, few members of the professional press use releases to generate stories. You won't see a press release reprinted in the *Wall Street Journal, Fortune, Financial Times, Business Week,* or *Forbes.* One reason is that the press receive thousands of releases in the mail every day. Members of the trade press use releases more often than members of the business press, but they too are being overwhelmed by the deluge of information hitting their desks. It simply isn't possible for journalists to read all of these releases. Most good journalists therefore rely on their contacts and industry word-of-mouth networks. The press also is highly competitive. No journalist wants to receive and write the same story in the same way as a competitor. Good journalists dig into the infrastructure of an industry for news.

The community. Every person who is interviewed by, delivers a package to, or visits a company walks away with an impression of that company. If company employees communicate properly, every person who comes in contact with the company becomes a salesperson for it—a carrier of good will about the company.

There are, however, problems with word-of-mouth campaigns. Sometimes it takes a while for a message to spread, and it takes a commitment of time by all the managers of the company to make word-of-mouth communication truly effective. But the benefits of word-of-mouth campaigns greatly outweigh the problems. You should learn how word of mouth operates in your business and industry. Then pass the word.

DEVELOPING THE INFRASTRUCTURE

In the mid-1970s, when personal computers first became available, most business people saw them as a passing fad. Not Ben Rosen. At the time, Rosen was working as an electronics industry analyst at the Wall Street firm of Morgan Stanley and was publishing a newsletter on the electronics business. He began writing, and talking, about personal computers. While others saw personal computers as toys, Rosen viewed them as the basis for a dynamic new industry. Through his newsletter articles and informal conversations, Rosen began to spread the gospel of personal computing.

Gradually, Rosen began to win converts. First, the readers of his newsletter became believers. Then *Forbes* magazine ran an article explaining how Rosen used a personal computer to perform financial analysis. More people started to pay attention to these new machines. Soon, companies began to view a favorable report from Ben Rosen as the key to success in the personal computer business. A bad review from Rosen was the kiss of death.

Why was Ben Rosen so influential? Had he been an obscure industry analyst, his endorsement of the personal computer, or any particular personal computer company, would not have been very significant. But Rosen was considered one of the top analysts in the country. He also had a command of the technology and the marketplace. People believed him to be a credible source of information.

Rosen played a central role in the infrastructure of the electronics industry. Every industry has an infrastructure, though it takes a somewhat different form in each case. The infrastructure includes all those people between the manufacturer and the customer who have an influence on the buying process and who contribute in some way to the complete customer solution. These people give credibility to products and companies (or, in Rosen's case, to a whole new industry). Without the support of the infrastructure, a product, service, or company is sure to fail.

I like to picture the infrastructure as an inverted pyramid, with the manufacturer at the bottom and the customers at the top. Figures 8 and 9

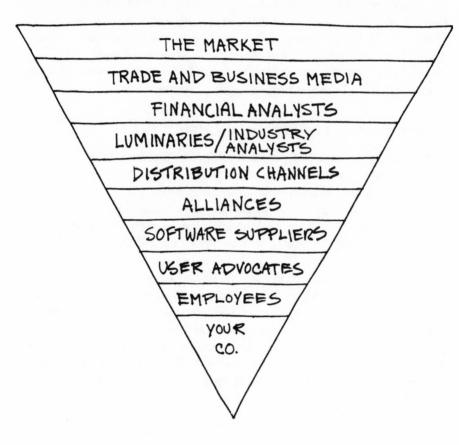

THE INFRASTRUCTURE DEVELOPMENT
OF THE PERSONAL COMPUTER
INDUSTRY

Figure 8.

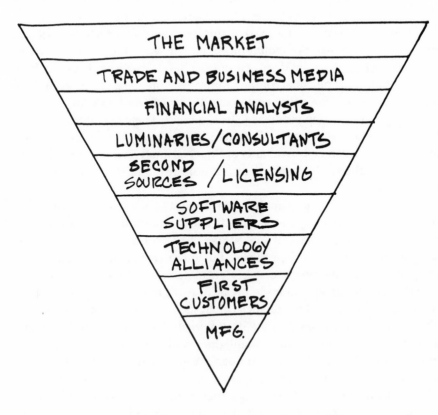

THE INFRASTRUCTURE DEVELOPMENT
OF THE MICROPROCESSOR
INDUSTRY

Figure 9.

show the infrastructures for two industries—those of personal computers and microprocessors. In each case, information about the product and the company bubbles up to the market through the infrastructure pyramid, largely through the word-of-mouth process discussed in the previous section.

Each level of the pyramid influences other levels, particularly those above it. Take a look at the personal computer infrastructure. Beta-site users or Early Adapters become referencing customers. If third-party hardware and software companies develop products to be used with a new personal computer, industry opinion leaders, or luminaries, begin to take notice. These luminaries, who can be consultants or financial analysts or key users, begin to influence others. Dealers and distributors respond to the views of the infrastructure.

Next come the financial analysts. An analyst's favorable report on a new company, product, or service can significantly improve that entity's chances of success. Where do financial analysts get their information? They get some from the manufacturers themselves. But more comes from resellers and distributors, systems integrators, key customers, and luminaries.

The business press and trade press, in turn, rely heavily on comments and recommendations of the financial analysts. Here's a test you can do yourself. Take any article from the *Wall Street Journal, Business Week, Forbes,* or any news publication. Then take a highlighter and mark all the quotes from sources other than the featured company. You probably will find that most of the source material is from analysts and sources within the infrastructure.

At RMI, we analyzed 180 corporate profiles appearing in major U. S. trade and business publications between August of 1988 and August of 1990. We wanted to learn the sources of the information on which the profiles were based. Here's what we found:

Number of Quotes (by Source)	
Financial analysts	84
Market analysts	70
Users	54
Other journalists	30
Third-party suppliers	29
Consultants	19
Competitors	14
Dealers	9
Other	37

The Context of the Quotes	
Corporate product strategy	155
Technology	93
Product	72
Market environment	72
Management	41
Financial matters	24

Totals are different because some individuals were quoted more than once.

It is interesting to note the significant role industry watchers play in the infrastructure. When influential luminaries tag a computer as a future leader in the marketplace, more software companies begin writing programs for the computer. Eventually, the word reaches the top of the pyramid and customers begin buying. The infrastructure development process is about getting allegiance to the product from important players. It also is about educating others, building alliances, and making deals with participating partners. Each part of the infrastructure validates the others and helps to build credibility for the product and the company.

The situation is similar with microprocessors. If a new microprocessor is supported with development tools, peripheral chips, and software, users are more comfortable in committing to that particular architecture. This is very important, as few customers will commit to a new microprocessor without a lot of visible market support.

If a company is missing any level of the infrastructure, the whole pyramid can come tumbling down. National Semiconductor ran into this problem with its 32-bit microprocessor. The product is excellent, technically superior to most of its better-selling competitors. But because it has few peripheral chips and little software support, the microprocessor has scored very few design wins. Without all the supportive tools in place, the reference system falls apart.

As a result, National was not able to build a reputation as a major force in the market. During the time National was trying to get its processor accepted by the market, I participated in a board meeting at a manufacturing company that was trying to decide which microprocessor to use in its next-generation products. A full three-quarters of the discussion focused on qualitative factors. Board members were looking for a microprocessor maker they could count on for new products and support in the future. Despite its highly advanced product, National was quickly knocked out of consideration. Why? It had no history of success in the microprocessor business and little support from the infrastructure.

The computer business today is a battle to create your own infrastructure. I call this battle "camp marketing." Apple, Sun, Microsoft, Hewlett-Packard, Digital, IBM, and Compaq are all trying to create their own camps. The leader of the biggest camp gets to set the standards for the industry and, to a certain degree, control the marketplace. But it is not possible to have just one company set the standards in such a technology-intensive, fast-changing industry. IBM did it with mainframes in the '60s and '70s but it took a combination of IBM, Intel, and Microsoft to do it with personal computers. Some consider Microsoft the dominating and controlling company in personal computer marketing today. Microsoft established its camp early and enlisted thousands of followers. But the technology is changing with the emergence of a new generation and, as in the past, such changes have resulted in a new cast of leaders. New camps are challenging the old, and old players are realigning to create a new leadership camp. For example, the new generation of desktop computers will be based on a computer architecture called RISC (Reduced Instruction Set Computing). Intel, Motorola, MIPS, and Sun (SPARC) all offer different versions of RISC architecture. IBM and Hewlett-Packard make their own proprietary versions. Apple, Compaq, and other computer makers will choose one of the available RISC architectures. The camp with the strongest following will no doubt create the biggest market. The stakes are high and the camps are busy lining up. Deciding which camp to be in is not a matter of going with superior technology performance. It helps. But the camps are formed for more strategic reasons. One executive told me that fear and personal enmity play a major role in the formation of various camps. Clearly a camp strategy can be won only by building relationships and developing the infrastructure.

The infrastructure tends to be particularly important in rapidly changing industries with complex products. In these industries there is so much going on that it is difficult for even knowledgeable people to sort out all the details. To understand the significance of new developments, people rely on what they hear from the infrastructure. No software application developer is going to commit money to a new program unless the infrastructure supports the computer platform on which his application will run.

In general, infrastructure marketing is becoming more valuable to a wide range of industries and services. This is happening because markets in all industries are becoming more segmented. I believe every industry has an infrastructure. In fact, since the publication of the first edition of this book I have had letters and calls from lawyers, elected officials, consumer-products marketing managers, military strategists, and even a pet-product producer, all talking about how the principles put forth in this book apply to their respective industries.

Infrastructure marketing may be more important for marketing computers than for stereos, and more important for stereos than for cereal. Pundits and analysts don't talk nearly as much about new cereals as they do about new computers. But each industry has some structured relationships that make that industry's market tick. The supporting infrastructure in Silicon Valley is the most sophisticated of any infrastructure outside of Wall Street. It is not a purely technical, engineer-to-engineer environment. The social–business structure consists of consultants, lawyers, venture capitalists, bankers, university officials, and accountants, as well as the technical players.

So how can a company line up a supporting infrastructure? I recently talked with a group of junior marketing people at a successful personal computer company. They told me about their plans to use public relations to create consumer demand for their product. I told them that their plans were bound to fail. Traditional public relations is not enough. Public relations can get your product mentioned in *Time* magazine once a year—if you are lucky. But that won't sustain much demand for a product.

Instead, marketing people must work at identifying and lining up the key members of the infrastructure—and keeping track of how the infrastructure is changing. In the computer business, they must identify a few highly visible beta sites, independent software people, systems integrators, industry consultants and analysts, luminaries, and key journalists, and then get all of them committed to the new product. They might try to start by providing on-line support for third-party software and hardware suppliers and dealers, or by running private briefings for industry luminaries. They might give special demonstrations and technical support to independent software companies.

All the time, marketing managers should pay attention to the hierarchies of influence existing within the infrastructure. For example, some luminaries are more "luminous" than others. When Intel introduced its first 16-bit microprocessor it gave an extensive briefing to Gordon Bell, a well-known technology guru, then director of engineering at Digital Equipment. When *Fortune* magazine ran an article about the new chip, it ran a quote from Bell to back up Intel's claims about the chip's likelihood of success.

Certain distributors and dealers are more influential than others as well. Higher up the hierarchy are specialized providers or value-added resellers.

Each market has a few high-quality firms. The important thing to remember is that your customer will judge you by the company you keep.

In the computer business, lining up the right software companies is even more important. The major system-software companies can make it easier for the smaller software companies to develop applications programs for the computer. And once the software companies commit themselves, manufacturers of add-on hardware, such as plug-in boards and disk drives, are sure to follow.

When a company is able to develop the infrastructure fully, it gives its product a big head start on the success trail. The product is a certain success even before it reaches the market. Perhaps the best example is Lotus 1-2-3, the integrated software product from Lotus Corporation. Once again, Ben Rosen played an influential role. Rosen, who now heads a venture-capital fund, was the primary investor in Lotus. He began talking about the product months before its introduction. Lots of people had early prototypes. I had one. Most computer producers had early prototypes. We talked to each other about it and the excitement grew. We could hardly wait for the final product to hit the market. By the time of introduction, 1-2-3 was the industry's worst-kept secret, but also its most sure-fire success.

FORMING STRATEGIC RELATIONSHIPS

Relationship marketing is essential in developing industry leadership, customer loyalty, and rapid acceptance of new products and services. Building strong and lasting relationships is hard work and difficult to sustain. But I believe that in a world where the customer has so many options, even in narrow product–market segments, a personal relationship is the only way to retain customer loyalty.

In fast-changing industries these relationships are becoming more important than ever before. As technologies advance and become intertwined with one another, no single company has the full range of skills and expertise needed to bring products and solutions to market in a timely and cost-effective way. To produce a personal computer, for instance, a company needs expertise in semiconductor technology, display technology, disk-drive technology, networking technology, software applications, communications and systems integration, as well as other areas. No company can keep pace in all of these areas by itself.

As a result, collaborative efforts are proliferating. Fast-growing companies, once fiercely independent, are now forming all sorts of alliances, even with former competitors. Every small company, it seems, is looking for "sponsors," while large companies are trying to link up with as many innovative start-ups as they can. *Business Week* magazine wrote in a 1984 special report on the computer industry: "For companies large and small, collaboration is the key to survival." These collaborations can take many forms: joint ventures, technology exchanges, manufacturing agreements, and equity positions, among others. Although some of the agreements seem aimed at R&D or finance, these relationships can play a critical role in a company's marketing strategy.

Companies in fast-changing industries need to form strategic relationships for a variety of reasons:

- To compete in today's markets, companies need a diverse set of technologies. Fields involving computers and communications are merging, and customers want complete solutions. No company can develop all of the necessary technologies and solutions by itself.

- The costs of developing new technologies are rising rapidly. Companies must share costs if they are to survive.

- Global competition requires that many of the old nationalistic trade issues be put aside in order to access and expand markets. U.S. companies are facing increasing competition from Japan, Germany, and Korea. Yet many U. S. companies are also forming alliances with non-U.S. companies in order to rapidly access markets and share the costs of capital to do so.

- Technologies are changing more quickly than ever before. At one time, a company could stay at the forefront of many different technologies. Now it is much more difficult.

- Small companies need to gain management expertise, distribution muscle, and capital in order to compete. Strategic relationships can provide these.

- Less tangible, but just as important, strategic relationships can bring added credibility to the companies involved. By choosing the right strategic partner, a company can gain credibility by association.

Many strategic relationships link a small company with a large company. These relationships are not a zero-sum game: both companies can benefit.

KEY RELATIONSHIP FACTORS

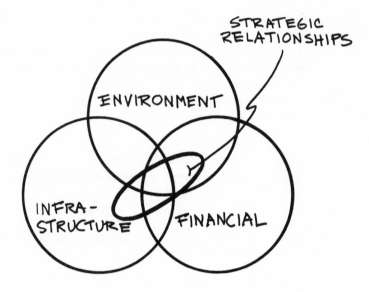

Figure 10.

A small, growing company acquires an important aura of credibility by linking up with a large, respected company. The large company acts as a credible reference that tells the market the small company is a winner. For the start-up or small company, a corporate partnership is a strong influence in raising additional capital from the venture capital community. Customers are more willing to take a chance with a small company if the small company has IBM, Sony, Intel, Microsoft, or Apple standing behind it.

At the same time, large companies can gain a window on new technology. Typically, small companies develop new technologies faster than large, bureaucratic companies. So by forming links with small companies, large companies can bring more innovative products to the market, and get them there quicker. Companies in the pharmaceutical industry have been doing this for years. It is a fast and relatively inexpensive way to develop new products and markets. Small companies simply do a better job of getting innovative, new products to market and of discovering new market opportunities.

A good example of this type of strategic relationship is the alliance between IBM and Microsoft. IBM agreed to use Microsoft's MS-DOS software as the primary operating system on its personal computer. The operating system, essentially the traffic cop controlling activity inside the computer, is a critical element in a computer system. Designers of the operating system and of the computer itself must work closely together. For that reason, IBM had always developed its own operating systems for its computers. But the deal with Microsoft made sense for both companies.

For Microsoft, the IBM deal meant instant credibility. Microsoft was an obscure company in Washington State, founded by Bill Gates, then only in his 20s. Suddenly, Microsoft was seen as a significant company in the personal computer industry. Its revenues have soared ever since, exceeding $1 billion in 1990. For IBM, the Microsoft deal meant the giant company could get its personal computer to the marketplace much faster than it could have otherwise. IBM was already somewhat late getting to the market. If it had to develop its own operating system, it might have arrived too late to become a leader.

IBM has forged other alliances as well. To help with the development of floppy disk drives, it struck a deal with Tandon. For microprocessors, it decided to standardize on Intel's family of 16-bit processors. It also invested in Intel, initially buying 12 percent of Intel's stock, and later increasing its stake to 20 percent. In each case, IBM gained quick access to new technology, while its smaller partner gained an important shot of capital and credibility. IBM's stamp of approval on the small company delivered a clear message: This company is a winner.

Some will argue that Microsoft benefited more from their agreement than did IBM. That may be true, but IBM was able to access the personal computer marketplace much faster in this situation than it would have had it made the move alone. It is estimated that personal computer business represents more than one-third of IBM's $70 billion in cumulative sales. It is my experience that small companies tend to benefit more from strategic relationships than do the bigger partners. American companies have a lot to learn about building and maintaining mutually beneficial alliances. The Japanese have found a way of working with big and small companies as well as with competitors in order to own a particular market.

The February 25, 1991, issue of the *Wall Street Journal* reported in an article headed "How Big Companies Are Joining Forces With Little Ones for Mutual Advantages" that "big companies like Glaxo are increasingly turning to close alliances with small companies...the small companies get money and credibility that a big corporate ally can provide. Though the entrepreneurs risk being smothered by giants, many are learning to protect themselves—and finding that they are in a stronger position than they once thought."

These strategic relationships should allow each company to maintain its independence and unique corporate character. These alliances should not be confused with traditional acquisition and diversification moves. Acquisition strategies often suppress innovation rather than foster it. The larger company often forces the acquired company into its corporate mold, thereby killing the innovative character of the small company that made the company attractive in the first place.

Indeed, acquisition strategies in technology-based industries have a pretty dismal record. Schlumberger, for example, tried to acquire its way into high technology. The company, a leader in the oil-services business, wanted to gain a foothold in new technologies, so it acquired Fairchild Semiconductor, the pioneer of Silicon Valley's semiconductor industry. But the strategy backfired. Key employees left the company and Schlumberger's corporate culture did not translate well to Silicon Valley. Schlumberger's desired foothold has turned into nothing more than a toehold. Exxon Corporation's effort to enter the office-automation market through acquisition of small high-technology companies turned into a disaster. Western Electric acquired robot maker Unimation in 1982, then saw Unimation's sales drop sharply. And AM International's high-technology acquisitions drove it into bankruptcy in 1982.

Pharmaceutical companies use alliances and acquisitions to broaden their portfolio of products and patents. They seem to have better success with this strategy than companies in other industries. Eli Lilly acquired Hybritech, a successful venture-backed company that pioneered monoclonal antibodies. Hoffman LaRoche acquired Genetech, the pioneer in biotechnology. Bristol Myers bought Genetic Systems and Johnson & Johnson bought Lifescan. Most of these acquisitions appear to be successful ventures. More recently, DuPont and Merck merged their R&D centers to form DuPont Merck.

Consumer industries are also addressing the need to acquire new products and processes by both building alliances and acquisition. Procter & Gamble is an equity-participating partner with Metaphor Computer, a computer company in Silicon Valley. Cathy Nichols Manning, the director of McKinsey's Los Angeles division and a leader of its Consumer Goods practice, says, in the Winter 1990 issue of the *McKinsey Quarterly*, "Today, many leading consumer companies are struggling to replicate historical successes. That is not proving easy in the new environment created by global competition, changing demography, flattening and fragmenting consumer demand, and growing retail trade power. In the packaged goods industry, for instance, which is normally thought to house much of the industry's marketing talent, product development has become largely ineffectual. Despite years of hefty R&D expenditures, there have been few genuine new product successes. In fact, industry participants are increasingly looking to acquire smaller companies that have been able to commercialize new ideas. Many now see acquisition as a more cost-effective way to introduce new products than internal development."

Clearly, acquisitions are filled with pitfalls. In many cases, large companies would be wiser to buy minority interests in small companies, or sign development contracts with them. These approaches allow the small companies to maintain their culture and entrepreneurial zeal.

To better understand the growing need for strategic relationships, it is important to understand the product-development cycle in technology-based businesses, shown in Figure 11. There are many steps between the scientist's workbench and the assembly line, and no company can handle all steps. Strategic relationships are needed to bridge the gaps.

As the figure shows, the product-creation process breaks down into four stages:

Basic research. Although most industrial advances rely on progress in basic research, industry funds very little of this work. Basic research seeks to answer fundamental scientific questions, such as what is the internal structure of matter, or what are the properties of human-body proteins. Research is a long and uncertain process. No one can know ahead of time whether it will lead to new applications or technologies. Few companies have the resources or patience to fund this type of research. So most basic research in the United States is performed at universities or national laboratories, with funding coming primarily from the government.

Applied research. When a scientific endeavor becomes directed toward a particular industrial result, it becomes applied research. Squeezed between basic research and development, applied research suffers the woes of a middle child: ambiguity and neglect. University researchers, whose main goal is to expand scientific knowledge, prefer to focus on basic research. Small companies can't afford to get involved until research has already passed through the applied stage. So most applied research is performed at large industrial labs, such as those at IBM, Hewlett-Packard, AT&T, Hughes, and at Xerox's Palo Alto Research Center.

Development.   This is the most directed phase of product creation. Its goal is a finished product that can compete in the marketplace. All companies do development work. But in fast-changing industries, small companies are the most productive and successful in development efforts.

Unlike corporate giants, which must invest substantial resources in maintaining their bureaucratic structures, small start-ups direct almost all their resources and energies toward development of a product. Necessity presses start-ups to be more innovative: the company's very survival depends on the success of the development effort.

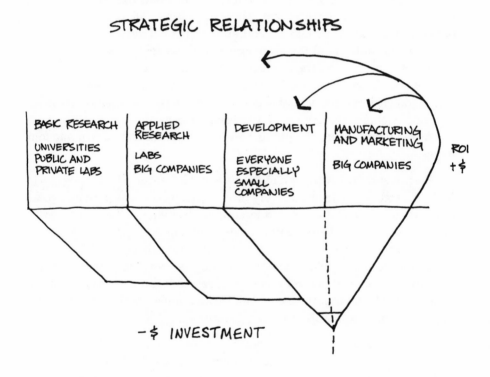

Figure 11.

Manufacturing and marketing. The first three stages of the cycle, from basic research through development, all represent investment costs. It is through manufacturing and marketing that companies produce a return and recoup the costs of product creation. Without this return the investment is lost and it has generated no new capital for the next generation of product innovation. Company size and resources are often major factors in manufacturing and marketing, so big companies tend to be the leaders in this final stage of the product-creation process.

It is in this final stage that Japanese companies have their biggest advantage. Japanese companies usually lag behind their U.S. counterparts in the first three stages of the product-creation cycle. But in some fast-growing markets—such as those of consumer electronics and semiconductor memories—they have managed to leap ahead in the final stage, partly because of superior manufacturing technologies and partly because of the special treatment they receive from the Japanese government and the Japanese financing system. In winning the manufacturing battle, the Japanese have deprived U.S. companies of the returns needed to invest in future-generation products.

Strategic alliances can help companies share costs and expertise, and thus pull together the resources necessary to have joint ownership of a market. In looking at the product-creation process, it is clear that different companies have different strengths in different parts of the process. Teaming up is a way to share expertise. If big companies typically are stronger in applied research and manufacturing, while small companies are the most innovative at development, why not join forces? IBM's strategic alliances have done just that, linking IBM's manufacturing prowess with the developmental skills of Intel, Microsoft, Tandon, and others. It is estimated that in the past five years IBM has established hundreds of relationships with much smaller companies.

Strategic relationships will become more important as global competition becomes more intensive. Time compression is creating a greater sense of urgency—a necessity to act rather than to plan—to be engaged in the technology and the marketplace, wherever it may be. In addition, the costs of research, product development, maintaining production facilities, and lost time are becoming burdens to even the wealthiest of enterprises.

These pressures are causing an increasing number of companies to explore ways of accessing technology and markets through cooperation. Competitors are even working together to meet the demands of their markets. Dawoo and Thompson CSF formed a joint venture in ceramics. Motorola and Toshiba formed product development, technology, and purchase agreements, and jointly formed Tohoku Semiconductor Corporation. Sierra, Singapore Semiconductor, and National Semiconductor joined together to develop, produce, and market advanced application-specific integrated circuits (ASICs). Many other companies have joined as well: Amdahl with Fujitsu, MIPS Computer with Kabota and Toshiba, Syquest with Nippon and Jafco, and Benzing with Kanematsu.

Businesses throughout the world are forming a global Silicon Valley: Chips come from Japan and Korea, assemblies from Singapore and Mexico, software from France and the United States. Multinational sourcing (acquiring desired products, components, or technologies), manufacturing, and marketing have become commonplace.

Currently, about 40 percent of U.S. trade involves one branch of a company selling to another branch located in another country. Modern companies source globally, using the newest technologies from around the world.

Increased international interdependence is changing trade in another important way: the national origin of products is becoming increasingly difficult to recognize. Because technology businesses organize their operations on a worldwide scale, their products cease to be truly American, or Korean, or Japanese. Parts, components, subsystems, products, and services are intermingled and exchanged. This trend is causing greater international economic integration and is moving us toward a true global economy.

The international nature of Sun Microsystems illustrates the globalization of business. Sun is a Silicon Valley venture-capital start-up established in 1982. It is one of two companies credited with creating the engineering workstation computer market. The company grew to $1.5 billion in its first seven years. Sun buys chips from Toshiba and NEC, uses Fujitsu to manufacture its next-generation miniprocessor, licenses workstation designs back to Toshiba, then acquires from other U.S. firms chips that are fabricated in the Philippines, Japan, Singapore, and Scotland. And throughout its operations, the company uses equipment made in Germany, Switzerland, Holland, and Japan.

The Ford Escort, the prototypical high-technology–produced car, is another example. The Escort is made in Great Britain by an American manufacturer, and contains parts from eleven different countries.

This pattern of global interdependence is typical and essential to the high-technology industry. As more countries join the technological trade arena, they must also join the global market system.

Small companies have to ask themselves whether they can survive without the resources and credibility offered by larger companies. For companies that start to slip, establishing strategic relationships will be the only way to regain credibility and to build a new image. These slipping companies need to make a dramatic change.

The idea of forming strategic relationships is not limited to electronics and computer businesses. It applies to all fast-changing industries. Strategic alliances can be critical in the biotechnology industry. As in the electronics industry, most of the innovation in the biotechnology industry comes from small firms. But bringing products to market is particularly difficult for small companies in biotechnology. Many biotechnology products must gain government regulatory approval. The wait for such approval is a long and expensive process. Few small companies have the resources to wait it out. Teaming up with large companies solves their problem, while also giving the start-ups much-needed marketing muscle and credibility.

Strategic relationships helped establish Genentech as a leader in the biotechnology industry. Genentech had a fair amount of credibility from its very beginning because its funding came from one of the most respected venture-capital firms: Kleiner Perkins Caufield & Byers. Using that initial credibility, Genentech was able to attract the interest of Eli Lilly, a pharmaceutical giant. The two companies signed a deal under which Genentech would develop human insulin using recombinant-DNA technology and Lilly would produce and market the product. This deal gave Genentech production and marketing capabilities it never could have financed on its own. Equally important, the association with Lilly gave the start-up an aura of credibility and established it as the technology leader in an infant industry.

While strategic relationships seem increasingly attractive, they are not without problems. Nothing is an automatic success. There are many factors working against the formation of strong bonds between companies. Perhaps the biggest source of problems between partners is poor communication. Oftentimes things don't get done because each partner believes the other is responsible. When entering into a new relationship, all companies involved need to be explicit about their objectives and expectations. The companies must agree on all details: what is to be done, by whom, and when. Management responsibilities and financial policies should be clearly stated. In some cases companies define their markets and goals very differently from each other. These types of philosophical differences should be aired and resolved before any agreement is reached.

Antitrust also can be a problem. In light of the economic challenge from Japan and other international competitors, the U.S. government is allowing companies a greater degree of flexibility in structuring alliances. The government has raised no objection to the Microelectronics and Computer Technology Corporation, an alliance of a dozen computer and electronics companies that was formed to share research costs. But other relationships are sure to raise objections. The *intent* of the partners is the critical factor. Relationships structured to restrict competition are, and should be, unacceptable.

Another problem is that small companies can become overly dependent upon their larger partners, much as military contractors survive at the whims of the Pentagon. Companies that depend on a single relationship to supply them with their primary source of business can end up in big trouble. MiniScribe, a tiny Colorado company that supplies disk drives to IBM, saw its stock plummet by more than one-third when IBM changed its buying patterns. The situation can be even worse when a large company decides to vertically integrate, developing its own production capabilities for parts that it once bought from outside partners. Small companies must remain aware of where they stand in their partner's plans, and never should get in a position where their very survival depends on the continuation of the relationship.

SELLING TO THE RIGHT CUSTOMERS

Customers are the key to any business. Companies are always looking to attract new customers. However, many companies fail to realize that *which* customers they attract is often more important than how many customers they attract.

Just as companies should look to form strategic relationships, they should try to sell to strategic customers. An impressive customer list can give a company a reputation as an innovator or a technological leader. Tandem, the pioneer in "nonstop" computers, sold one of its first systems to Citibank in New York. To outsiders, the message was clear. If Citibank trusted Tandem, then Tandem must be a winner. *Business Week* then quickly ran an article on Tandem, based in part on the Citibank reference. If Tandem had sold to Southwest Mutual as its first customer, it probably would have taken much longer to develop its reputation.

Key customers can help in other ways too. They can give valuable feedback, providing a company with new ideas on how to improve a product. What is more, key customers feed information about the manufacturer into the word-of-mouth network. If every key customer tells two other people about the company, and each of them tells two others. . . . You get the picture.

Companies should pay attention to choosing the right customers even before they introduce their product. Picking the right beta sites to test early versions of a product can be critical to the product's ultimate success. Valid Logic, a new manufacturer of computer-aided engineering systems for the electronics industry, found that one of its beta sites, Convex Computer Corporation, provided important suggestions for improving the product. Convex, itself a high-powered start-up, was developing a highly sophisticated supercomputer, and thus was able to push the Valid Logic system to its limits. At first, the Convex engineers complained about the Valid system. But as Valid responded to the criticisms and modified its prototype, Convex became an important Valid supporter.

Convex itself faced an important decision in choosing beta sites for its supercomputer. I had long discussions with Convex about the decision. Should the beta site be at a university? How about a government agency? Or a military contractor? Each possible site had its own unique characteristics. Convex would learn different things from different sites. Equally important, the choice of a site would start to position the company. By choosing a military contractor, Convex might be positioned as a military supplier, and then find it difficult to sell to the commercial market. Ultimately, Convex chose two beta sites: a semiconductor company that is using the computer for chip design, and a petroleum company that is using the computer for geophysical research. Identifying and selecting the right customers in each of its important markets has been one of the key strategic marketing elements for Convex. The references from customers in each segment have helped the company build itself into one of the two leading manufacturers of supercomputers.

Deciding which customers to sell to requires creative segmentation of the market. Many companies selling industrial products give segmentation little thought. They segment the market geographically, or into big and small companies. This is especially true of start-ups. Ask a start-up: What's your market? The answer is predictable: The Fortune 500. They all think that selling to the Fortune 500 is some type of magic formula for success.

Well, marketing doesn't work that way. Fortune 500 companies are large, bureaucratic organizations. They have numerous rules and qualification criteria that products they buy have to meet, and they generally are hesitant to try new technologies and products. Selling to those companies is a long and tedious process, lasting a year or more. Start-ups would be better off selling to the "emerging Fortune 500"; that is, the 500 companies most likely to grow and be successful in the next decade. These companies have to make purchase decisions more quickly, they are more likely to try new and innovative products, and they probably will come back for repeat purchases as they grow.

Finding the emerging Fortune 500 is not necessarily easy. Tomorrow's winners are not always readily apparent. Predicting which of today's companies will be tomorrow's successes is not done purely statistically; it's done more qualitatively. We can't just tear a chart out of *Fortune* magazine. But identifying these companies can pay off. ASK Computer, in Mountain View, California, has put this type of strategy to work. ASK sells software to manufacturing companies. Rather than initially targeting the Fortune 500, it first sold to other high-technology companies in Silicon Valley. Its customer list includes many of the fastest-growing companies in the country. Because these companies don't drag their feet on purchase decisions, ASK was able to grow quickly to a $100 million company.

Another, closely related, approach to segmenting the market involves the "adaptation sequence." Social-science researchers have noted that people can be divided into four categories according to how quickly they adopt new products and beliefs. Some people lead the way. They are the Innovators. Next come the Early Adopters, then the Majority. Finally there are the Laggards, who are the slowest to adopt new ideas. According to one book on the subject, about 2.5 percent of the public are Innovators, 13.5 percent are Early Adopters, and 16 percent are Laggards.

These groupings can be used to classify companies as well. Companies have attitudes just as people do, and these attitudes can be used in positioning new products. Only one change: I like to think of companies as *adapting* to new technologies, rather than *adopting* them. That is why I call the process the adaptation sequence.

Figure 12 breaks down the personal computer market into Innovators, Early Adapters, Late Adapters, and Laggards. Members of different groups have very different motivations and attitudes. Innovators are fascinated with technology and are willing to educate themselves about new products. Laggards, at the other extreme, are much less knowledgeable about new technologies and will not purchase a new type of computer unless they have an absolute need for it. They respond to competitive pressures. Selling to these different groups requires very different strategies.

There is a great temptation to target Laggards in your marketing strategy. First of all, there generally are many more Laggards in the marketplace than there are Innovators and Early Adapters. What is more, Laggards typically are large corporations that could lend immediate credibility to your business.

The Market Adaptation Sequence

Adaptation Criteria	Small Business/Personal Computer			
	Innovators	Early Adapters	Late Adapters	Laggards
Product Acceptance	Technology Fascination	The Coming Thing	Obvious Solutions to Problem	Absolute Need
Motivation	Implement New Idea	Leap Frog Competition Improve Business	Competitive/Social Pressure Fear of Obsolescence	Extreme Competition/ Social Pressure
Confidence Level	Willing to Experiment High Self Confidence High Risk	Willing to Try New Things Will Go with Reasonable Risk	No Risk Slow to Change Needs References	Reluctant to Change Culture Problems Strong Justification
Education Attitude	Self Taught Independent	Will Attend Night School to Learn	Will Attend Seminar Wants to Buy a Proven Product Needs Lots of Hand Holding	Will Send Someone to Seminar Needs Proof Ease of Use
Acceptance Criteria	Latest Technology New Features Performance	Innovation Better Way to Do Job Selective	Brand Important Pay for Only Needed Features Terms and Conditions Important	Lowest Cost Competitive Terms and Conditions Brand Very Important
Selling Strategy	Self-Sold, Once Turned On Word of Mouth	Benefits Reference Word of Mouth	Address Cost Problems/ Technical Support Needs Examples Demonstration	Productivity Increases Fear

Figure 12.

But it usually makes more sense to aim new products at the Innovators and Early Adapters. Innovators are more likely to take a chance with a new product or new technology. And because Innovators are usually, though not always, small companies, they make purchase decisions relatively quickly. Moreover, the actions of Innovators influence all the others. Innovators spread information about the product through word of mouth. If Innovators buy a product, others are likely to follow suit. So selling to an Innovator might actually bring more credibility than selling to a Laggard, even if the Laggard is many times larger and better known. As Innovators influence companies downstream in the adaptation sequence, credibility for the product grows.

Let's look at an example. General Electric is a major Intel customer, and I would classify it as a Late Adapter. GE made a commitment to use Intel's 8086 microprocessor in a variety of its products. But when I talked to a senior manager at GE in 1982, he was getting nervous about the decision. He noted that many innovative companies in Silicon Valley, including Apple and Fortune Systems, had decided to go with Motorola's 68000 microprocessor. Intel's 8086 still had dominant market share, and its customer list included many major companies. But the Motorola chip took on a certain aura. The GE executive said to me: "Those start-up companies in Silicon Valley are capturing our imagination. If we had to make the decision today, we would not go with the 8086."

Large companies can sometimes afford to wait until Laggards begin to buy a product. Then, they can sell in large volume to the Laggards. But small companies can't afford to wait for the Laggards to come around. They must target the Innovators, preferably the most visible of the Innovators. If I were a start-up selling disk drives and I could choose to sell to Apple, Xerox, AT&T, or a start-up like Metaphor, I probably would choose Apple and Metaphor. Their orders might be small, but my reputation would be established. Rather than spend millions of dollars on promotion, I could simply rely on word of mouth.

Of course, identifying the Innovators in an industry is not always easy. Not all Innovators are small. Some pockets of large organizations are Innovators. At banking giant Citicorp, for example, the MIS division is an Innovator. At General Electric, some divisions are Innovators, while others are Laggards. It is necessary to understand the organizational forces acting within large companies in order to decide whether to target certain divisions. Intel has an Innovator's Program that identifies which divisions at other companies are Innovators and which are not. Other companies might benefit from similar programs.

Propaganda ends where dialogue begins.

Marshall McLuhan

Chapter 7

Communications: From Monologue to Dialogue

PS ʸ

There are a variety of ways to communicate with the infrastructure. Some forms carry more weight than others and some are more expensive than others. There are direct and indirect forms of communication. Direct forms include telemarketing and face-to-face sales. Indirect forms of communication use intermediaries to deliver the messages. I consider advertising an indirect form of communication. An ad is a monologue, not a dialogue. Ad readers can't ask questions, or nod yes or no. Ads are information, not communication. Direct mail falls into the same category. It is difficult to judge the value of indirect communication.

Communication is a dialogue. When communication is effective both parties gain something. Communicating with customers in the new marketing involves listening as much as talking. It is through dialogue that relationships are built and products are conceived, adapted, and accepted. As all markets begin to look like niche markets with clear infrastructures, the communication process becomes more and more direct. In almost all forms of business, seminars, users-group meetings, workshops, industry trade shows, customer and dealer advisory councils, and boards and trade associations are proliferating. Companies are learning the value of direct communication.

Marketing was easier years ago when markets were more monolithic, few companies competed, and buying tastes were much more uniform. The producer controlled the market. With fewer producers and products, and more shared consumer tastes, mass advertising worked. In today's environment, "other" has become a major player in almost all markets, from fast foods and cookies to personal computers and semiconductors. Brand loyalty is dying. Consumers are more willing to try the new name brand. Witness the rise of clones in the computer business or of Mrs. Fields cookies. Most industries fragmented, rather than consolidated, in the 1980s. There is more of everything available to the consumer. That can be both good and bad.

With so many products in narrow categories, consumers are often confused. But even with all the confusion, no evidence shows consumers scrambling back to the security of established brands. With thousands of products and brands trying to gain the customers' attention and loyalty, consumers have gained the power of choice. And as the technology allows more businesses to create options, consumers will be in an even more powerful position, because options feed choice. How, then, can businesses keep customers loyal? They must establish relationships with their customers. This doesn't mean that the president has to shake hands with every customer. That's not possible. Relationships are established through experiences. Experiences are a communication medium.

Let me give you an example of what I mean. I have had a number of good experiences at The Good Guys, a chain of stores in California that carry consumer electronic products. A year or so ago I purchased a Panasonic portable disc player. I have no idea whether Panasonic disc players are the best. A salesperson who frequently helps me demonstrated all the models, and based on his knowledge, I bought the Panasonic. Three days after I bought the player I dropped it on my foot. It didn't appear broken but it would not work. Since I still had the box, and even the Styrofoam protective materials, I rewrapped the player and took it back to the store.

On the way to the store I began to wonder how much trouble I was going to have getting the machine either replaced or repaired. (I once had a similar experience with a television set and it took three months to get it repaired.) Because the player did not look broken, and the packaging materials were intact, I could have told the salesman that the product didn't work when I first opened the package. What do you think most consumers would do? I decided to see what would happen when I told the truth. I told the salesman I dropped the player on my foot and broke it. He went to the shelf and handed me a new machine. I was surprised. I know The Good Guys were probably protected by the manufacturer, but I was overwhelmed. The no-questions-asked policy allows customers to communicate honestly with the dealer. On another occasion, the same salesman helped me find a professional-model tape recorder, obtained literature on the product from the manufacturer, and recommended a dealer who specialized in high-end products. I not only tell this story over and over to friends and business associates, but I buy only from The Good Guys regardless of price. These experiences taught me to trust that company. They taught me that I could communicate openly with the salespeople and that these people would help me. Service is good communications more than anything else.

Managements must begin to think how they can create positive relationship experiences with their customers. Salespeople, dealers and distributors, detail people, telephone operators, service and support people, third-party consultants—and all others who touch the customer—are communication mediums.

The other side of this coin is the feedback mechanism necessary to complete the dialogue. The American automobile producers got into trouble because they believed that the sale was complete when the dealer bought the cars. While the channel is an important intermediary, it is not the ultimate customer. Feedback must be obtained from the product's users. In this fast-changing environment, buying habits and tastes change quickly. Tom Peters of *In Search of Excellence* fame has long championed customer involvement in the product-development and manufacturing processes. This can be accomplished by establishing advisory groups and panels or by making it easy for customers to call and ask questions, suggest solutions to problems, or give feedback. Tandem Computers, producers of on-line transaction processing computers, invites client companies to brainstorming sessions. By combining Tandem management's ideas with client brainpower, new uses for the computers that will help customers gain a competitive edge in their industries are created. And by integrating its machines into clients' strategic plans, Tandem can better plan and develop products and services in the future. Software companies are on the leading edge of setting up telephone and on-line systems to give customers and users access to the company. Because of the nature of their product, software companies must provide a convenient way for customers to get help. All companies should consider on-line services or an 800 number. These communications tools can be used to provide relationship-building opportunities.

Computer technology is dramatically changing the way communications is done in business today. Computer networks tie workers more closely together than any company newsletter can. Sun Microsystems president Scott McNealy said that "the network is the computer." As networks proliferate, businesses will be linked directly to customers and every member of their infrastructure. Interactive communications is rapidly becoming a reality. The dialogue between producer and supplier and all the intermediaries has already begun in the computer industry. With interactive computing technology, for example, a product called Bookseller's Assistant eliminates the need for a salesperson by helping customers narrow their choices for reading material. Yet the system is no less personalized than a clerk's assistance. A browser enters into the system titles of favorite movies, books, and magazines. Based on that information, the computer prints out a customized list of books that person would most likely enjoy. A revamped, formerly low-tech trucking industry offers customers computerized access to shipments. Skyway Freight Systems of Santa Cruz tracks shipments via computer, then lets customers tie into the company's systems in order to monitor their own shipments without middleman delay. Interactive technology also offers a computerized just-in-time (JIT) delivery system similar to Skyway's system that has allowed Japanese manufacturers to keep tight control of shipping and inventory costs. JIT shipping allows companies to hold down shipping and inventory costs by insuring that a shipment gets to the manufacturing plant precisely when the production process demands it.

Sitting at my computer in my home office, I can communicate with every employee in my firm, all the partners at Kleiner Perkins Caufield & Byers, my wife in her office, my broker and bank, all of the people at Apple Computer, as well as their dealers and third-party support organizations, Oracle Corporation, and dozens of other clients. The computer will have as profound an effect on the way we market products as the airplane, car, and television have had on other aspects of our lives.

Data bases that house easily accessible information about customers and markets are already changing the identification of smaller-market segmentation. When this information is seen as a marketing tool and a way of gathering customer feedback, and not just a way to identify and sell a product, good marketing will begin.

Marketing communication is much broader than advertising, direct mail, or public relations. While they have varying degrees of value, based on the product complexity and the market size, they can't create customer experience. Building a communications strategy and process based on word of mouth and insuring a positive customer experience is hard work. It's a lot easier to run an ad, issue a press release, or hold a press conference. It's a lot easier to talk than to listen. But in the age of the customer, we have to learn that communications are as important as the product.

Market Relations™. Because the press reflects what the infrastructure says about your company and your products, traditional public relations approaches don't work. Traditional public relations work focuses attention on getting "good ink." Stories and ideas certainly can be "sold" to the press. After all, the press need a lot of material to fill the pages of regularly scheduled publications. So the push–pull mechanism creates lots of ink. But your company can't get good ink without the references of your infrastructure. Managers at a large European computer company complained to me that, although it was one of the first companies to introduce the 386 personal computer, it got no recognition in the American press. Why would it when it had no distribution strategy, no major customers to reference, and no agreements with third-party software suppliers? It was absent from the infrastructure. Market Relations™ is the development of communications programs at all levels of the infrastructure. Its purpose is to create the alliances, references, and word-of-mouth communication necessary for a company to capture a leadership position. You can't achieve a sustained and positive public relations program without first having a sound Market Relations™ program.

The first task of positioning a product is to prepare the infrastructure. Most U.S. journalists practice what I call he-said-you-said journalism. Rather than present their own analysis of a situation, they simply quote what other people say. And who are these other people? Most often they are members of the industry infrastructure—customers, financial analysts, consultants, distributors, and resellers. The infrastructure serves as a filtering mechanism, helping journalists separate fact from fiction.

Companies should take advantage of this filtering mechanism. They should educate and win over members of the infrastructure before going to the press. If a company tries to go to the *New York Times* or the *Wall Street Journal* without first developing the infrastructure, it could run into big problems. Reporters will go to members of the infrastructure for information about the company and its products, and the company might not be happy with what the people in the infrastructure say.

Market Relations™ is a process that takes time and effort. However, it has been my experience that the most effective way to sustain a market position is to sustain market communications. If companies would spend as much time and money on meeting and talking with key customers, field sales managers, key resellers, and key third-party support organizations as they do on advertising or promotions, sustained positioning would be a lot easier.

Matter and antimatter. It seems that in our present environment every position, story, or approach has its supporters as well as its highly visible detractors. I call this support and opposition matter and antimatter.

Charles Ferguson, an MIT political scientist and electronics-industry consultant, argues that big firms, like IBM and Intel (for which he has worked), are the only ones with enough staying power to compete against Japan's large, stable, technologically progressive semiconductor companies. Start-up companies that raid talent and technology from larger firms, he says, sap the strength of the industry, while acting as mere piddling players in the long-term global economic scene. To combat this situation, he urges government to tax capital gains in order to discourage investment in fledgling entrepreneurial ventures.

On the other side of the issue is George Gilder, a conservative writer whose book *Wealth and Poverty* influenced Reaganomics. He argues that young, competitive companies are precisely what keeps the industry as a whole vital and innovative. Innovation doesn't come from the sclerotic big guys; it's small companies, armed with talent, private or venture-capital backing, and zeal to develop products in the most efficient way possible, that represent the unfettered free market at its best.

If a company is going to merge, stockholders come forward to both praise and damn the management. Scientists argue about the impact of the deteriorating ozone layer and whether there is a global warming trend. Environmentalists urge fast-food chains to abandon Styrofoam containers while others protest the use of paper products because such use requires the destruction of forests. Claims of new and advanced technology are met with counterclaims that receive more attention than the original claim. Members of the electronics industry trying to raise consortium are shouted down by the anticonsortium people. Issues are analyzed, discussed, debated, and researched until something else distracts press attention. Pro and anti groups form around every issue. Little wonder that managements have grown fearful of dealing directly with the press. Almost every public subject now has a legal dimension. Pharmaceutical companies have a major concern with health and safety issues. There is even a growing concern over the radiation effects of computer screens on the health and safety of workers. Every company will experience its share of antimatter. Managements must begin to take communications more seriously. They must participate in developing strategy, anticipating problems, building credibility, and communicating directly with the infrastructure. When public issues are involved, the infrastructure may include local, state, and federal government officials and agencies, public-interest groups, and the editorial board of industry publications.

A recent example of a company looking bad in the face of antimatter was shown on the television program "20/20". The featured investigative story was about a small pharmaceutical company that produced collagen products for use in cosmetic surgery. Although thousands of people have used the company's products without ill effects, some people apparently have adverse reactions. Several women who had been injected with the company's collagen material (a substance that occurs naturally in the human body) had severe reactions to the material. Claims were made that the material caused extensive physical debilitation and even life-threatening complications. The company produced its testing data and medical references. But "20/20" produced its own medical researchers who said the company's research was either poorly done, irrelevant, or incomplete. As each antimatter witness spoke, an impressive, authoritative title was displayed on the screen. Most viewers certainly would not be able to judge the credentials of the antimatter spokespeople. It was clear from the statements made by these witnesses that the company did not deal with the complaints of patients and physicians who had problems with the products. When the president of the pharmaceutical company was questioned by "20/20", he was faced with antimatter from the start. He became so flustered he ended the interview. Antimatter had won. Antimatter can't be dealt with in public. It must be overcome by establishing direct and honest communications with customers.

Most managements are comfortable dealing with trade publications because these publications serve specific industries and rarely run reports that are based on speculation. These publications generally employ journalists and reporters who are knowledgeable about the particular industry and are focused on products and services. As management's concerns broaden to financial, trade, environmental, and public issues, media concerns intensify. In today's society issues are complex and have many seemingly paradoxical elements. Add this to the growing litigious nature of our society and it's a wonder anyone wants to be public.

Daniel Boorstin, noted American historian, wrote an insightful book about the media and the rise of antimatter in our society entitled *The Image: A Guide to Pseudo Events in America.* He says, "In a democratic society like ours—and more especially on a highly literate, wealthy, competitive, and technologically advanced society—the people can be flooded by pseudo-events. For us, freedom of speech and the freedom of the press and of broadcasting includes freedom to create pseudo-events. Competing politicians, competing newsmen, and competing news media contest in this creation. They vie with one another in offering attractive, 'informative' accounts and images of the world. They are free to speculate on the facts, to bring new facts into being, to demand answers to their own contrived questions. Our 'free market place of ideas' is a place where people are confronted by competing pseudo-events and are allowed to judge among them. When we speak of 'informing' the people this is what we really mean."

Journalists are interpreters of information. When you speak through the press, you are not talking directly to your customers. The interpreters of information don't have the same objectives you have, nor should they. Your objective is to have your products or services and company viewed in a positive light so that customers are predisposed to buying your products. Unfortunately, controversy and antimatter attract readers. It is difficult to communicate your message when there is so much extraneous information and antimatter overwhelming the media. This is the reason so many companies are producing their own journals. Sun, Microsoft, Lotus, Oracle, IBM, and Hewlett-Packard all have their own publications which they use to communicate directly with their customers. There is no generic answer to the question of how to deal with antimatter. Opposition will always be a part of our society. Managements must take it seriously and deal with it constructively by developing more direct communication channels with their customers and their infrastructure.

Communication is management's role.   At most companies, top management pays attention to press coverage only when it delivers a negative message about the company or when the competition gets a lot of positive coverage. Communications functions are delegated to lower-level employees. A recent quote in a major business publication read, "The company spokesperson said, 'No comment.'" Top managers should play a more active role in public-communications strategy. Communication is what relationship building is all about.

At most companies, especially small, technology-based companies, the personality and culture of the company can be traced to the management team. As the company grows and marketing plans proliferate, corporate personality often fades. Top managers become the only ones able to communicate the corporate character and ideals. They are the only ones who can offer a simple, unified view of the total corporation.

Communications must involve all of the company's employees. Companies spend huge sums of money announcing a product to the media and analysts but rarely hold formal announcements for their employees, even though every employee touches the customer in some way. The shipping clerk who has a clear understanding of the product, the position the company is trying to achieve, or the urgency of the marketplace is more likely to do a better job.

I know managements of billion-dollar companies that perceive communications as a promotional event rather than a process. "Get the message across," is the rallying cry. "Hold an employee meeting or press conference."

Such events seldom are truly effective in convincing people of a particular viewpoint. Tom Peters's *In Search of Excellence* spoke about "walk around management" and its effectiveness. Well, "walk around management" must also walk outside the company—walk and talk with customers, walk and talk with resellers, walk and talk with suppliers. Communications is a process, not an event. Communications is relationship building. Communications is as much listening as it is talking.

If layers of people are put between company management and key players in the infrastructure, the marketplace never will get a true sense of what drives the company. If, on the other hand, top managers meet regularly with key members of the infrastructure, journalists, financial analysts, and employees, everyone would benefit. Each group would come away with a better understanding of the others' positions. There would be less likelihood of misunderstanding and distrust and surprise. Cultivating communications is a job no public-relations agency can do for a company.

DEALING WITH THE PRESS

In my infrastructure model, I place the press near the top of the pyramid. It is my view that the press reflects the views and opinions of the movers and shakers within the infrastructure. It is essential to build the base of the pyramid before talking to the press. Too many companies think press relations come first. They want to make a splash in the press even before they build their reference structure. They think a good article in *Business Week* or *Fortune* or the *Wall Street Journal* can create a demand for their products and build their reputation. They believe a strong media campaign can cut the time to success, and make up for deficiencies in product quality, customer relations, and other basic marketing skills.

But press relations can't change reality. Press relations don't create a company's image; they reflect a company's image. Press relations can't take the place of a broad-based marketing strategy. Companies first must concentrate on establishing themselves in the infrastructure and building favorable references. Then, and only then, should they worry about getting press coverage.

All managements fear "bad press." They don't appreciate competitors getting better press than they do, either. The old idea that "knowing someone" in the press will lead to a glowing story just doesn't happen, at least not with the quality media.

Good press relations can be a valuable asset. The press can help reinforce and broaden the credibility the product and company have already gained. It can educate, and can ease fears by making customers feel secure about new technologies.

In new and fast-growing industries, journalists can play the role of evangelists. They can preach the new technology. The large number of computer enthusiasts among the press corps certainly has helped expand the markets for personal computers.

Advertising can perform many of the same functions. But information coming from the press is usually more credible. Articles in the media are perceived as being more objective than advertisements. If a company can win favorable press coverage its message is more likely to be absorbed and believed.

Building successful press relations requires time, planning, and constant reinforcement. It rests with an understanding of how journalists work and how information is communicated. I have put together a set of guidelines useful in developing an effective press-relations strategy.

Don't go to the press too early. Obviously, no company wants negative press coverage of a product before the product is even introduced. But positive coverage can be almost as bad. A favorable article appearing while the product is still in development might build expectations that are difficult to meet. If problems crop up and slow the development cycle, as so often happens, attention becomes focused on the delay rather than on the product.

Synapse Computer ran into this problem. Synapse had impressive credentials. The company, started by a group of engineers from Data General, planned to build "fault-tolerant" computers that never would break down. It had an excellent chance to succeed in its targeted market. But managers had the itch to announce that it was great before it actually was. Even while still working in the back of a candle factory, the Synapse management began running ads and talking to the press. Infrastructure expectations rose. Then, Synapse's computer ran into technical problems at the beta sites. The product did not perform well. But by then Synapse was very visible. Journalists were watching Synapse, and they reported on the company's problems. Synapse's credibility suffered. Regardless of whether it solved the technical problems, Synapse faced an uphill struggle, that of regaining its credibility. The company did not survive the challenge.

One of the main complaints Americans make is that a product just advertised isn't yet available. Attention focused on a product before it is available or before it is adequately distributed may create sales for its competitors.

Their interest aroused, consumers may well purchase another product that is available. First customers are the most important asset to building a referral base. Management should make sure those early customers are happy before they talk to the press.

Meet with journalists one on one.   Many companies build their press strategies around press releases and press conferences. But these are not the most effective modes by which to communicate a message. National magazines get thousands of press releases every week. It's tough for a company or product to stand out from all that information. And many press releases are thrown out without even being read. Nor are press conferences very effective. There are two problems with them. First, journalists are reluctant to ask their best questions at press conferences, because they don't want to alert the competition to their angle on the story. Second, different parts of the media have different interests. *Byte* magazine wants to hear about nanoseconds and megaflops. The news magazines want the broad trends and social implications. It's impossible to satisfy everybody. There's a lot of information, but not much good communication. A press conference is a nice spectacle, but the press loses out—and so does the company.

Instead, companies should meet members of the press individually. A one-on-one meeting takes time, but it makes more of an impression on the journalist and it delivers the message more efficiently than a press conference. Messages can be tailored for the audience: one for the trade magazines, another for the business magazines, a third for the general-interest press. Once again, the 90/10 rule applies: 10 percent of the press influence the other 90 percent. So select the most influential members of the press and meet with them.

Educate the media.   Maintaining press relations should be seen as an education process. Fast-moving industries are becoming more diverse, fragmented, complex, and difficult to understand. At the same time, there is more information available about every facet of every industry. For most journalists, these industries are becoming more and more confusing.

Companies need to help journalists create order out of the chaos, so journalists can present a cogent description of emerging trends and technologies. Rather than simply pitching ideas to the press, managements must be willing to spend the time to educate the press. Companies should treat journalists as well as they treat their major customers. It's not enough to hold up a new product at a press conference and say "Here it is."

Look beyond products. In new industries, the press typically focuses on products. The stories are generally naive and superficial. Most of the coverage comes from the trade press. But as an industry matures, so does press coverage. Journalists learn, question, dig into the "news behind the news." The business and general-news media become increasingly interested.

Companies must deal with the business and general-news media differently than they deal with the trade press. These media must place much less emphasis on product performance and characteristics. Seasoned journalists know a technological advantage is short-lived. Companies should explain how they fit into present and future business environments. Products should be discussed in terms of how they affect things like "global competition," or "white-collar productivity," or "manufacturing technology." The press is fascinated by glimpses of what lies ahead. As I discussed before, the environment defines the product.

Be honest about bad news. When bad news strikes, it's not worth fighting the press over whether to release it and exactly what to say. As a politician once told me: "Never pick a fight with someone who buys his ink by the barrel." Being honest scores points with the press. In unfavorable situations a company's character and style will greatly influence how the press perceives and writes about the company.

When there is bad news it is best to get it out. In our society I don't think it's possible to hide bad news. Trying to hide the news causes it to fester and linger. The press has a long memory, even if the public doesn't. Take the accident at Three Mile Island as an example. The Nuclear Regulatory Commission withheld information, and public confidence in the NRC, as well as in the safety of nuclear power, sank. On the other hand, Johnson & Johnson was very open with journalists during the Tylenol scare, and Tylenol has since regained its credibility in the market.

But the most effective way to deal with bad press is to deal with your customers first. When things go wrong, talk first to your employees and then your customers. If the people closest to the problem or most affected by the problem are informed and satisfied with your response, the press will reflect that attitude.

PUTTING IT ALL TOGETHER

Several years ago a well-known industrialist told me that all business success is based on two things: building relationships and patience.

Nowhere is this more true than in market positioning. Not one of the market-positioning activities—using word of mouth, developing the infra- structure, forming strategic relationships, selling to the right customers, developing communications—will guarantee success by itself. And none will bring success overnight. It takes a long time to establish contacts and build relationships.

But taken together, and given enough time, these elements are almost certain to work. They will bring recognition and credibility to a company and its products. Business is about people, not things. When a company concentrates its business on building market relationships, it can't lose.

When you're rich, they think you really know.

Tevye, from *Fiddler on the Roof*

Chapter 8

Corporate Positioning: There's Only One Thing that Really Counts

Pg 48

A strong corporate position is hard to achieve and even harder to regain. Just as product positioning gives individual products a unique presence in the market, corporate positioning provides a unique presence for an entire company.

Corporate positioning is based on many factors, including management strengths, corporate history, and even the personalities of the top executives. A well-known entrepreneur can help position a start-up company. People such as the late Bob Noyce, founder of Intel, and Bill Hewlett and Dave Packard give their companies a unique personality. Noyce co-invented the integrated circuit and was highly respected in engineering circles, so it was much easier for Intel to position itself as a technology leader.

But by far the most important factor in corporate positioning is financial success. Without financial success, everything else is meaningless. A company without profits will not maintain its position for very long. When people buy complex products such as computers or telecommunications equipment, they are making a long-term commitment. They don't want to buy from a company with financial troubles or an uncertain future.

In the musical *Fiddler On the Roof* the main character, Tevye, sings a song entitled, "If I Were a Rich Man." If he were rich, he sings, he would sit in the synagogue and lecture to the important men in town. "And it won't make one bit of difference if I answer right or wrong. When you're rich, they think you really know." When purchasing technology and complex products, consumers want to buy from a winner. The buyers use financial performance as an indicator of the company's product acceptance and success. The financial condition of the company sends signals of stability, or instability, to the marketplace.

On a plane trip between Boston and Washington, DC, I once sat with the president of large computer company in the midst of an unfriendly takeover. He had nearly missed the plane. Collapsing into the seat next to me, he asked the stewardess for a glass of water. He looked like he had been working long hours. The toughest part of dealing with the threat, he said, was all the travel he had to do to assure customers that the company was still viable. Acquiring new business, he said, was almost impossible. The flip side of the argument is also true. People feel more secure when they buy complex products from a company with a strong balance sheet. Everyone is eager to listen to profitable companies.

At a board meeting I attended, the manufacturing manager was reviewing the company's suppliers. He said that he asked the president of one of his semiconductor suppliers—a billion-dollar company—to come in and discuss that company's recent earnings loss. When one of the other board members pointed out that the supplier had plenty of cash in the bank, the manufacturing manager said it didn't matter. "When companies lose money, they cut operations, product lines, people, and service. I want to know what they have planned. I can't afford to have my production disrupted."

Corporate positioning sits at the top of the positioning hierarchy. Companies must first position their products. Next, the products must gain infrastructure acceptance. The result of being successful in these two areas is a strong corporate position.

As the last element in the positioning trio, corporate positioning reinforces the previous two. A strong corporate position can validate a company's market positioning and its product positioning. When a company establishes a strong corporate position, its other positions become stronger and more lasting.

The Japanese effort in marketing RAM (random-access memory) chips illustrates this point. First, Japanese semiconductor companies made the decision to position their products as high-quality parts. They invested in manufacturing operations that minimized defects in the chips. Their early penetration of the U.S. market was based largely on offering cheaper prices than offered by U.S. suppliers. To reposition themselves, Japanese suppliers selected a few target customers, such as Hewlett-Packard and IBM, and did additional product testing to make sure the chips going to those customers were top-notch.

The Japanese gained market position when Hewlett-Packard began running quality-comparison tests using American and Japanese chips. The Japanese chips showed fewer defects in test after test, and HP announced the results publicly. In effect, HP held up its scorecard and showed the Japanese as winners.

Had strong relationships existed between U.S. semiconductor companies and their customers, U.S. companies would have known about the quality differences much earlier, and I doubt Hewlett-Packard would have made its findings public.

After the HP tests, Japanese RAM chips gained market share quickly. Some customers checked other types of Japanese chips and found that they, too, were high quality. Soon, Japanese semiconductor companies had a solid corporate position as high-quality suppliers.

Sometimes a corporate position can be established on the basis of just one or two key products, which I call "silver bullets." Silver bullets are particularly important in technology-based businesses. If a company chooses its silver bullets carefully, and gains highly visible acceptance by marketing them, it can gain a strong corporate reputation, even if the rest of its product line is only mediocre. In this case, image becomes the reality.

Products at the forefront of technology are difficult to mass produce. Typically, they sell at high prices and in low volumes. Companies make most of their money on volume items, but these products are usually "plain vanilla" products; that is, they are rarely at the cutting edge of technology. Companies sacrifice some product performance when they design a product for volume production. Moreover, it usually takes a while to work out all the production problems. By the time the volume-produced product finally reaches the market, new products, using even more advanced technology, may already be out.

The ideal, then, is for companies to develop a balanced product mix. They should develop silver bullets to build an image of technological leadership. But to keep the money rolling in, they also should develop products that extend their existing product line or that can be used in combination with their other products. Take Xerox as an example. It probably will be some time before Xerox's new DocuTech Production Publisher, a sophisticated electronic publishing machine, brings in as much money as Xerox's broad line of copiers. But the DocuTech is a marvelous piece of technology and has received wide praise from customers and the media. This product has helped Xerox build its image as a technology leader. That image is critical to Xerox's corporate positioning. Xerox's image as a technology leader is a comfort factor. It reassures customers that they are buying from a company that will bring the latest technology to bear on solving their problems.

If a company manages to establish a strong corporate position, or corporate personality, it can reap many benefits. Corporate positioning tends to have long-lasting effects. Consider the case of the Japanese chip manufacturers described earlier. American semiconductor companies have improved on quality, and their products now match Japanese products in quality tests. But many customers still believe the Japanese make the highest-quality chips.

Among the other benefits of an established corporate position are:

Faster market penetration. A recognized strong corporate position attracts good partners. Everyone, from investment bankers to new start-ups, wants to deal with established leaders. Customers feel secure dealing with industry leaders.

Less product drag. Not all of any company's products will be world-beaters. A strong corporate position enables a company to be successful in selling its weaker products as well as its strong ones. A strong position can help a company hold on to customers when competitors challenge its products with new products.

Better access to market and technology information. Everyone wants a relationship with a leader. Everyone wants to in some way work with, or show their new products to, discuss their new ideas with, or sell to a leader. Companies like Sun Microsystems, Microsoft, and IBM do a very good job of listening to new potential partners.

Lower cost of sales. When a company has a strong corporate position, the market accepts the company's new products more readily simply because they carry the company's name. The company's salespeople don't have to work as hard.

Higher prices. Companies with strong corporate positions can sometimes charge higher prices for their products or services. Often they can set the umbrella pricing in an industry.

Better recruiting. Leading companies can recruit the best talent, because people want to work where the "action" is.

More employee loyalty. A strong corporate character encourages employees to identify with the company's success. It provides focus and direction throughout the organization.

Higher price-earnings ratio.   Investors are attracted to companies with a strong corporate position.

Of course, all these benefits will disappear if the company's profits begin to slip. As mentioned earlier, profitability is the most important factor in corporate positioning. The minute profits decline, the market begins to worry. Everything else is called into question.

Tandem Computers provides an example. Tandem was the first company to sell nonstop computers—computers that never break down. Since it shipped its first computer in 1976, Tandem started to build a reputation as a high-quality company. Its technology was advanced and its market positioning was strong.

But when Tandem's earnings dropped below expectations in early 1984, analysts began looking for reasons. They decided that Tandem's market positioning and technology were slipping. They wrote reports about new start-ups that were putting pressure on Tandem. Tandem's hard-earned positioning began to slip away.

The truth is that Tandem's technology and products were just as reliable as ever. The new start-ups were not a threat. Tandem's problems were more complex, relating to forecasting and pricing. The problems were soon corrected, but the damage to Tandem had already been done. Financial performance clearly had a major impact on Tandem's marketing efforts.

Trilogy Systems ran into even worse problems with its corporate positioning. The company was founded by Gene Amdahl, one of the geniuses of the computer industry. Amdahl has an impressive track record both as an engineer and as an entrepreneur. Amdahl Corporation, the company he founded in 1970, has reached sales of $462 million.

So when Trilogy announced plans to develop a line of powerful computers using a new semiconductor technology called waferscale fabrication, it was able to quickly establish itself as a technology leader. Raising money was no problem at all, and the company received lots of favorable publicity. But its position was fragile. When the company began to have financial problems in 1984, its credibility disappeared. The members of the infrastructure began to question whether even the talented Mr. Amdahl could succeed once again. But the company folded.

People don't worry much about the financial stability of consumer-goods companies. People never question the financial stability of the company from which they buy a tube of toothpaste or a box of detergent. Most consumers don't even know what company makes the toothpaste they use. But buying an expensive technology-based product involves a greater commitment. If the company goes bankrupt, who will provide the service? Who will provide the new parts? Once a computer company, such as Unisys, Wang, or Data General, gets into financial trouble, it has a hard time holding on to its customers. No one knows whether the company will still exist in six months.

When a company loses its corporate positioning, the only thing it can do is to start the whole positioning process over again. It must go back and reinforce its product positioning, and then reestablish its market positioning. Finally, it can try to recapture its corporate position.

Intel went through this process in the recession of 1981–82. Profits fell at all semiconductor companies, and some companies even lost money. Intel had been a darling of Wall Street for ten years, but people began to wonder about the company. Many analysts questioned whether Intel's profit margins would ever return to their pre-recession levels. They pointed to the loss of key employees and the company's declining market share in the microprocessor market. They even began to question Intel's management practices. People no longer thought of Intel as an innovator or a technology leader.

Intel responded by focusing on products and product positioning. The company introduced more than 100 new products in 1981, a record for the company. Gradually, Intel shifted its focus back to technology and products. When the recession ended, profits returned and people once again began to view Intel as the industry leader.

This focus on financial success is likely to intensify, if anything. Using computerized data bases, people can look up financial information instantly. A factory manager who is going to buy a sophisticated piece of equipment can hit a button on a computer and get a complete financial profile of the company selling the equipment. If the company looks a bit shaky financially, the manager is likely to buy from a competitor.

Financial officers should keep this scenario in mind. Positioning strategy is not just for marketing managers. It is for all managers. Financial results can destroy a company's positioning—or it can solidify the company's position as an industry leader.

Sustaining advantage requires continuous improvement and change, not a static solution in which strategy can be set and forgotten.

Michael Porter

| Chapter 9 | # Developing a Strategy: Knowledge Marketing and Experience Marketing |

KNOWING WHERE YOU'RE GOING

Throughout most of the 1980s, the perception of what it takes to succeed was fed by a few highly successful companies—Apple, Intel, Compaq, Sun, and Microsoft. The vision was clear. Just starting a company was a badge of courage and the entrepreneur became an instant hero. It was the decade of the get-rich-quick entrepreneur. The entrepreneur was to be the savior of American business. The Europeans and the Japanese did not have entrepreneurial cultures and because of this they were expected to lose their leadership positions. Radical changes in business brought on by the information revolution, the dissolution of mass markets, social change, deregulation, blurring industry lines, international competition, and the advancement of technology all played to the strengths of small companies, while they wreaked havoc on rigid, slow-to-change large businesses. But businesses had to think small. In the minds of many at the time, there was a recipe for success in Silicon Valley. It was believed that a company should design innovative, user-friendly products using leading-edge technology; that it should offer customers a total solution to their needs; and that it should target Fortune 500 companies as its primary customers. Mix all these ingredients, it was thought, and sales should grow to $500 million in five years.

By the end of the decade, these beliefs had changed and a heavy dose of reality set in. Large companies didn't crumble. In fact, many are adapting to the new environment. Japan and European countries, particularly Germany, continue to dominate old and new markets despite the lack of venture capital and an entrepreneurial culture. The business world has become more complex and difficult. The nature of these changes is illustrated in my comparison of the Entrepreneurial Dream 1983 and 1990. I drew my observations about the changing entrepreneurial dream from the dozens of new business plans I review each year.

These changes illustrate the dynamic nature of technology-based businesses. Unfortunately, introducing a new product or building a new company isn't the least bit simple. Of course, advanced technology and sound management are more important than ever. But in today's highly competitive marketplace, unless a company can achieve a unique position, it will not succeed. Developing a strong positioning strategy is the key to marketing success.

How can a company develop such a strategy? We have already looked at the basic ideas behind dynamic positioning. Now it's time to look at how companies can put those ideas to work. How can a company identify the proper niches for its products? How can it gauge the trends and attitudes of the market? How can it convert those trends into a successful strategy?

The Entrepreneurial Dream		
	1983	**1991–1992**
Market:	Fortune 500 companies	Niches
Potential size:	$500 million in five years	$50 million in five years
Expected image:	IBM-like	Sony-like
Products:	User-friendly	Portable
Design of products:	Innovative	Graphic
Technology:	Leading edge	Standard/open
Contribution:	Total solution	Piece of solution
Industry growth:	Dynamic	Developing
Competition:	None perceived	Everyone
Why now:	Window of opportunity	Out of a job
Marketing:	Aggressive, worldwide	Channels
Backing:	Kleiner/Perkins, H&Q, Sequoia, Mayfield, A. Rock, Venrock, Seven/Rosen	Kleiner/Perkins, Sequoia, Mayfield, IBM, Apple, HP, Sony

	1983	**1991–1992**
Valuation:	$100 million	$10 million
Headquarters:	Silicon Valley	Silicon Valley
Management:	Veteran, experienced	Veteran
When founded:	January 1983	January 1990
Founder:	Ex-Intel	Ex-Apple
CEO/President:	Ex-Intel	Search underway
VP engineering:	Ex-Hewlett-Packard	Ex-IBM
VP marketing:	Ex-IBM	Search underway
Total employment:	Three	Eight
Manufacturing:	Low cost	Subcontracted
Strengths:	Depth of management	Technical niche
Weaknesses:	None perceived	Management

As I see it, developing a positioning strategy is a three-step process. To start with, a company must have a good understanding of itself—its strengths and weaknesses, its goals and dreams. Top managers should have a coherent vision of the culture and goals of the company. If different managers have widely differing visions, the company never will be able to develop a solid positioning strategy.

Second, the company needs to understand the market environment. That is trickier than it might seem. Most companies gather statistics about customer behavior. Then, they make decisions based on the market data. This quantittive approach is quite satisfying to numbers-happy MBAs. In most cases, however, it obscures reality. Instead, companies should use more qualitative approaches to understanding the environment. Marketing managers must develop an intuitive feel for the market. Rather than gather numbers, companies should listen to customers' needs, problems, frustrations, and desires. Customer comments won't translate into graphs, but the company that listens to them will come to a better understanding of customers and the marketplace.

Finally, the company must use this information to decide on a positioning strategy. There is no single formula for deciding on a strategy. Just as the world is filled with a tremendous variety of technologies and products, so too is it filled with a variety of positioning strategies. Every company must find its own road to success. Managers must keep an open mind and seek a variety of opinions before settling on a strategy. Then, once the strategy is in place, managers should be willing to adjust the strategy as market conditions change.

Let's go on to discuss each of these three elements of the positioning process: understanding your own company, understanding the environment, and, finally, deciding on a positioning strategy.

INTERNAL AUDITS: KNOWLEDGE MARKETING

It is remarkable how many people have trouble answering the simple question "What business are you in?" One of my colleagues interviewed seven people at a start-up company in Silicon Valley several years ago. She asked that seemingly simple question and got seven different answers. One person described the company in terms of product applications. Another described it in terms of the technology used in the products. Yet another talked about the nature of the marketplace. People always define the vision and direction of their company by the role they play within it. In this case, the descriptions conflicted.

The company was developing a good product. But the company had not developed a coherent vision. Does this really matter? It certainly does. If company executives don't have a uniform and clear understanding of what business they are in and where they are heading, they are likely to run into conflicts. As the market environment shifts, the company might not recognize that it, too, needs to change. Just as bad, company executives could find themselves at odds with one another over how the company should adjust to market shifts. Infighting could paralyze the company, leaving it behind while the industry speeds ahead.

Harvard professor Theodore Levitt presented a classic example in his article "Marketing Myopia," published in the *Harvard Business Review*. Levitt describes the plight of the railroads. He says that had the giant railroads defined themselves as "transportation companies" they might have adjusted more effectively to the coming of airplanes. As it was, the railroads saw the emergence of air transportation as competition, not as a new opportunity.

Similar situations abound today. Indeed, maintaining a clear but flexible view of a company's mission is more important than ever in these times of rapid change. A few years ago a Silicon Valley company tried to market computer-music software programs. The company went bankrupt when the market did not grow as quickly as expected. The company's problem was that it saw itself as a computer-music company. Had it seen its mission as producing "creative uses of computers," or even "computer-based entertainment," it would have had a better chance of success. It could have adjusted more effectively to the changing environment.

Semiconductor companies Texas Instruments, Intel, and National Semiconductor all entered the consumer-electronics business in the early 1970s by producing electronic watches, games, and calculators. All of them failed. The consumer-electronics business was completely foreign to semiconductor managements. Manufacturing technologies, distribution systems, product design, advertising, and merchandising were completely different than in the semiconductor business. When Hewlett-Packard entered the handheld calculator business, it succeeded by making products that served engineers, its traditional customers. The vision of the company, which is an articulation of "what business you are in," must become the culture of the company.

Much has been written about corporate cultures in recent years, and culture can indeed be a powerful force. Each company must understand the internal cultural factors that drive the company. If a company can develop a culture that emphasizes quality and reliability, or service and customer relations, all employees are likely to work hard to see that those things are delivered. Similarly, a company with a culture that encourages innovation is more likely to develop creative new products.

In new companies, the founding entrepreneurs play a dominant role in determining the company's culture. Long after having given up day-to-day control of the company, Bill Hewlett and Dave Packard continue to have an important effect on Hewlett-Packard through the cultural norms they established at the company. Founders should take time to consider the cultural qualities they want their companies to have. If several founders are involved, they must make sure their ideas of corporate culture are in sync.

Understanding the internal forces at a company should come naturally, but often it doesn't. Many managers are so focused on the outside forces affecting their companies—new technology, new competitors, new markets—that they forget to look inside. They fail to examine the process of how decision making takes place.

When we begin to work with a new client at RMI, the first thing we do is force the company to do some soul-searching. We have developed a formal process called an "internal audit" during which we probe the inner workings of the company to find out what really makes it tick, and to make sure that different parts of the company aren't working against one another.

Typically, we meet with five or six top people in the company. We always talk to the president and the vice president of marketing, and we usually meet with the vice presidents of engineering, finance, and manufacturing as well. Each interview lasts for ninety minutes or so. During that time, we ask general questions about the company's history, its products, market, competition, and goals. We also ask the individuals about themselves, and how they perceive their role in the company. We perform these audits in order to compare the vision, goals, and objectives of the management with the perceptions we will gather later from customers and various members of the infrastructure.

Perceptions and attitudes held by company executives are critical to the company's positioning. Management must implement the positioning strategy and communicate it to employees, customers, and marketing partners. By examining the perceptions and attitudes of individual managers, we can expose inconsistencies and conflicts among company executives, and we can get a feel for the company's expectations. After finishing the internal audit we typically conduct an "external audit" of key industry observers and potential customers to determine whether the company's goals and expectations are, in fact, realistic.

Obviously, we vary the questions we ask in an internal audit from one company to another, and to a lesser extent, from one individual to another within the same company. You can get some idea of the process from the following questions, used during the internal audit at a software company:

- What business are you in?

- What are your company's fundamentals?

- Describe your market. What makes it tick? Who are the key players?

- Describe your company's technical, financial, cultural, and other strengths and weaknesses? What are your customers' views of your strengths and weaknesses?

- Who are your competitors? What are their strengths and weaknesses?

- How competitive are your products? What would you change to make them more competitive?

- What are the key milestones and timetables that must be met in the next two years?

- How do you segment your markets? What are the key factors for success in each of your market segments?

- What are the significant trends in each of these market segments?

- What is your pricing strategy?

- What is your distribution strategy? How will you support each market segment?

- How important is service to your company? What is your service strategy?

- What *must* you change in order to become or remain the leader in your marketplace?

- What percentage of the company's resources will be devoted to each market segment? Are the resources adequate?

We aren't gathering simple facts and statistics during the internal audit. We're looking for qualitative information, such as management's perceptions and feelings. We try to avoid making any assumptions ahead of time, and we lead clients only to stimulate their own thinking. Oftentimes we will let clients ramble in order to learn more about their hidden feelings. Most important, we want to find out whether all the key members of the management team are singing the same song, and how deeply they believe in that song.

Even with simple questions, an internal audit can expose important information and insights about the company. Take, for example, the final question on the list. In estimating the percentage of resources that should go to the development of language software, executives at the software company gave consistent answers. All the responses indicated between 20 percent and 30 percent. But for operating systems, responses indicated from 25 percent to 75 percent. Estimates for consumer-market software also varied widely. These differing estimates exposed an important split in the company. A more qualitative analysis of the executives' comments backed up this finding. The company seemed to be split into two camps that in many ways were sabotaging each other. One camp, led by a manager who favored consumer marketing in the style of Procter & Gamble, wanted the company to focus on entertainment software. The other, led by a manager with a strong technical background, wanted the company to dedicate its efforts to producing systems software, which had a smaller market, but this market was one in which the company had already proved itself.

An internal audit at another company, this one specializing in communications, unearthed a different type of conflict. We found that the engineer in charge of software development thought he was the top engineering officer at the company. That would have been fine, except that the engineer in charge of hardware development thought the same thing.

Most companies, especially small and fast-growing ones, like to sweep these conflicts under the rug. They ignore conflicts until company performance starts to sag. That is a formula for trouble. It is critical that all members of the management team understand their own role in the company, and that they all share a common vision of the company's plans and goals.

Older, established companies find it difficult to communicate a common vision even when they know they must change in order to compete in a new environment. Two companies I worked with in the past two years were noticing dramatic shifts in the technology and the competitive environment. They had to change or fall behind. Both wanted to create a new direction and clearer vision for the future of the company. Both felt they had to get their management moving in the same direction. Yet both found the challenge too difficult, because the new vision created too many conflicts with managers of older operations. Building a consensus among management to support leading-edge investments and changes for the benefit of the whole company may be the most difficult task in a changing business environment.

To develop a common vision, different parts of a company must understand each other's needs and goals. The management team must understand and appreciate marketplace dynamics and customer needs. Everyone in the organization must participate in the relationship-marketing process. In particular, marketing and product-development staffs need to work closely together and learn to understand each other. At many technology-based companies these staffs hardly interact at all. Marketing people develop positioning strategies without even consulting the product designers, and certainly without having any deep understanding of the product-development process. This compartmentalization within companies is a serious problem. Each department builds walls around itself. Time to market is so critical in fast-paced organizations that seamless communication and responsibility sharing must become part of every business' culture. No one has a monopoly on a good idea or a better way of doing business.

Getting managers to cooperate in these ways, however, is a difficult job, with objectives that sometimes seem incompatible. Marketing managers must let the customer know the product is technologically advanced and exciting. They must not denude the product of its technology. People like to have the latest technology. That's why it's so important to have the technical people participate in the marketing and sales process. These people can best represent the advances and potential of the technology. When quality, reliability, cost, and service are concerns important to customers, manufacturing and service become significant sales and marketing functions. The role of marketing people, then, is to integrate the whole product for the customer.

The marketing role in technology is often one of translation. Most people involved in the buying-decision process of a complex product have no idea what the technology of that product is all about. Millions of personal computers are sold each year to people who want the latest technology but have no idea what the difference is between a 386- and a 68040-based machine. Yet, ads and promotions use these numbers to represent the latest technology. Intel's Ed Gelbach once called this the "hexachlorophene syndrome." In the 1950s a toothpaste maker advertised a secret ingredient called hexachlorophene, even though few consumers understood the term. Today, customers are excited to own a "16-bit processor," even if they don't know the difference between a bit and a byte. At the same time, customers want to be reassured about the new technology. Marketing managers must find ways to make customers feel comfortable with new ideas and new technologies.

To translate complex new technologies into appropriate symbols and messages, marketing managers must understand and appreciate the technological process. Without that type of understanding it is impossible to create marketing messages that capture the importance and meaning of the product.

Designers themselves can offer important marketing insights. In fact, I would say I have gotten more good ideas from technical people than from marketing people. Technical people have a lot of intuitive knowledge that doesn't come across on the spec sheet. Some engineers are not very articulate, but others can speak about technology in an almost poetic way. They have lived with the product and understand the technology on a deep level. They can explain why the technology is important, and where it is heading.

Companies whose marketing people don't mix with designers often miss out on great opportunities. The marketing people don't really understand what their company's products and technologies are all about. Xerox fell into this trap and the company has suffered as a result. Researchers at the company's Palo Alto Research Center had been at the forefront of the personal computing revolution. They developed many of the technologies that Apple used in the Macintosh computer. But Xerox marketing managers never had an appreciation for these advanced technologies. Xerox introduced many of the technologies—such as the mouse and bit-mapped display—in its Star computer two years before the emergence of the Macintosh, but Xerox's marketing managers did not position the product well. They didn't realize what a powerful technology they had. Several frustrated Xerox designers defected to Apple, and helped Apple turn those same technologies into a mass-market success.

EXTERNAL AUDITS: EXPERIENCE MARKETING

No company can develop a positioning strategy in a vacuum. No matter how advanced its technology, no matter how low its prices, no matter how extensive its distribution, a company will succeed only if it understands the market environment into which it is selling its product. That is, it must understand the strengths and weaknesses of its competitors, the perceptions and attitudes of potential customers, and the social and political trends of the market. A product that flopped miserably might have been a wild success had it been introduced six months sooner—or six months later. It all depends on the market environment.

Companies must satisfy customer needs, not simply produce goods. And to do that, they must monitor and understand the environment. For companies in fast-changing industries, this task is particularly difficult and important, as the environment is on a roller coaster of change. Only with constant and creative monitoring of the environment can these companies position their products effectively.

Statistics Don't Reflect Reality in the Marketplace

Research is often mistaken for actual marketing. If a lot of research is presented, the marketing people are thought to be doing their jobs. Research is considered the thing to do for anyone whose job title has the word "marketing"in it. But most research is abstract.

What is the value of market research? That all depends on who's doing it and how they go about it, as well as who is interpreting it. Most research simply fails to provide the insight and understanding necessary for management to make comfortable decisions. "Marketing research just hasn't delivered," say Jim Figura, vice president of research for Colgate–Palmolive Co., in an article in *Business Week* titled "The 'Bloodbath' in Market Research." "It is an embarrassment."

When the presidents of successful high-tech companies want to know what is going on in their markets, they get on an airplane and visit their customers. After making two or three visits, the presidents know what is going on. Getting as many managers out of the shop and talking to customers is the best research method a company can employ. It often seems to me that the conclusions reached after performing a lot of market research could have been reached by using common sense.

There are several reasons not to trust market research:

1. The environment in which the research is done is different from the actual decision-making environment. Advertising research is often a process of reviewing and analyzing a selection of ads. This is not how consumers make judgments about products and services. When asked to comment about ads, research subjects respond as critics, not as typical consumers.

2. Researching future products is impossible. Until faced with actually having to make the decision to buy a product, customers are making pre-judgments based on a perfectly defined world. Those judgments most often change over time, and are affected by price, available competition, and the economy.

3. Statistics don't lie, but they don't understand the customer either. Research is often used as a substitute for understanding customers.

4. Trends and changes can be understood only historically. Research is really about history, not about the future. Predicting the direction technology, the market, or customers may take is an art, not a science. To make such predictions requires being immersed in the marketplace or technology. It takes a judgment process based on experience.

5. Communication is so instantaneous and information so prolific on any given event that attitudes and opinions change on a daily basis.

Most companies survey the market to determine the level of customer demand. But qualitative aspects of the market are more important. By firsthand monitoring, a company can learn how receptive customers are to change and what "mental obstacles" customers have that the company must overcome to get people to accept new products and technologies. Through firsthand monitoring, a company can get a better sense of customers' expectations, their level of understanding, and their willingness to be educated. And it can act or respond quickly.

A few years ago, a manufacturer of digitally controlled equipment found that workers in mature process industries, like the tobacco industry, felt uneasy about new digital equipment. Why? Walking around the factory floors, the company's marketing people found that the workers were accustomed to operating equipment with knobs and gears. To operate the new digital equipment, workers had to push buttons, and they felt uncomfortable pushing buttons. They would rather turn knobs as they always had. So the manufacturer began installing knobs rather than buttons on its equipment. By discovering the problem firsthand, the manufacturer was able to correct it and thus did not lose customers

Management expert Peter Drucker has noted that companies can't really answer the question "What business am I in?" until they understand customer attitudes and perceptions. He explains: "What is our business is not determined by the producer, but by the consumer. It is not defined by the company's name, statutes, or articles of incorporation, but by the want the customer satisfies when he buys a product or service. The question can therefore be answered only by looking at the business from the outside, from the point of view of the customer and the market."

In an attempt to understand customer wants and needs, most companies perform market research and statistical analyses. In some mature, slow-changing businesses these techniques work adequately. Tire manufacturers, for example, have developed statistical-demand analyses that link tire sales to automobile sales. And toothpaste manufacturers can forecast demand largely by studying demographic data showing the number of people in each age category.

However, these statistical techniques don't work well in fast-changing industries that are moving into unexplored territory. When your company is creating new markets, no one really knows where it is headed. You have to be creative. John Sculley, chief executive at Apple, says he is wary of numbers-oriented analysis. He explains: "The only quantitative data I use are what people have done, not what they are going to do. No great marketing decisions have ever been made on quantitative data."

Indeed, market statistics are rarely meaningful in rapid-change industries. Look, for example, at the personal computer software industry. Projecting the size of this industry involves little more than guesswork. In 1983 three respected market-research organizations tried to project growth for the personal computer software industry. The three companies couldn't even agree on the size of the market when they looked backward, to 1982. When they tried to project the industry size for 1987, the numbers they came up with ranged from $3.7 billion to $13.6 billion. The industry size turned out to be more than double the amount predicted by most forecasters. These projections didn't stop Apple, Compaq, or IBM from investing in the industry's technology and market development. What can be concluded from numbers? Not very much. Numbers can be manipulated so much that you can choose any numbers you want, and reach any conclusions you want.

This case is hardly unique. In growth markets, few projections have ever been correct. Even the Semiconductor Industry Association, which gathers data directly from the companies in the semiconductor industry, is usually way off target with its projections. Surveying customers doesn't help much. Customers are always enthusiastic about future products, but they don't necessarily buy the products once they are introduced. In many cases, there seems to be infinite demand for the unavailable.

Nevertheless, market projections proliferate. Like economists, market researchers don't seem shamed by their repeated failures. Numbers seem to make people feel more secure, so everyone tries to come up with some statistics, however meaningless they may be. Rare is the marketing vice president with enough confidence to go into a meeting and say: "There simply aren't any good numbers available." Instead, marketing managers somehow create numbers that justify their plans.

Numbers with little scientific basis swirl around the industry and sometimes come full circle. When *Business Week* did an article on commercial uses of superminicomputers, in 1979, there were no market projections. So a *Business Week* reporter called three companies and in thirty minutes produced a graph for the magazine. A few weeks later the reporter was working on an article about computer printers. A leading manufacturer of printers showed him the *Business Week* graph and used it to justify his company's strategy and production plans.

The problem is caused, in part, by the way business schools train MBA students. Students are taught to rely on abstract theories and numbers-oriented planning. They have little appreciation for how fast-changing industries really work. Tektronix, a leading manufacturer of electronic instrumentation, learned this lesson the hard way several years ago. Recognizing that it needed to become more marketing-oriented, Tektronix hired a group of new MBAs with interests in marketing. These MBAs came to Tektronix armed with charts and theories. But they had no understanding of the peculiarities of the electronic-instrumentation business, and they never made much of an effort to learn. Within a few years nearly all the hot-shot recruits were gone. Tektronix still needed marketing help, but it learned that a flock of MBAs was not the answer.

The case-study approach used at many business schools causes additional problems. In the case-study approach, students learn by analogy. But there are no analogies to use for many new markets. The rules for marketing personal computers aren't the same rules as those used for marketing stereos or consumer electronics. They aren't even the same rules as those applied to marketing big computers. The marketing textbook has to be rewritten every day for emerging industries. Business schools should establish apprenticeships at companies, involve the businesses in the curriculum, hold workshops at the companies involving the students and company managers, and encourage new ways of looking at marketing.

Most traditional approaches to marketing try to accurately predict the unpredictable. Unfortunately, you can measure only what you can control. And no one can control how people will respond to new developments. People themselves don't know how they will respond to new developments. Politicians use poll results to guide their thinking on everything. Rather than act on conviction and leadership, they let the polls guide their policy decisions. As a result, we don't develop long-term solutions or sustain the commitment necessary to solve our problems.

Trying to make predictions about technology-based products is even tougher. In 1976 very few people thought they needed a personal computer. Traditional market research would have shown a market for a few hundred, not a few million, units. Only as companies brought new and useful software to the market did people realize a personal computer could, in fact, be a useful tool. Recognizing this problem, some leading-edge companies pay no attention to market research. James Levy, former president of Activision, a software company, explains that "market research will kill as many good games as bad ones. It's not a definitive tool. You never know about a new title for sure until it hits the street."

Levy believes companies in fast-changing industries must wean themselves from numbers. "People want everything to be predictable," he explains. "They try to turn everything into a science. They're uncomfortable with uncertainty. But to be successful in this business you must deal with and live with uncertainty and surprises. A certain amount of uncertainty you must accept as a fact of life."

Indeed, it is insecurity that drives the development of new markets. Insecurity keeps companies sensitive to changes. Statistical information, with its illusion of certainty, provides a false sense of security. Companies relying on statistical data become lax and less sensitive to market changes. In fast-changing markets a company can't afford that.

THE QUALITATIVE APPROACH

If marketing managers can't rely on numbers, how can they ever understand the market environment? They must use a combination of intuition and keen sense for changing attitudes. They must develop a feel for the market, just as a card player develops a feel for the table. That doesn't sound as scientific as statistical analysis, but the fact is, it works better.

Developing a feel for the market, an intuitive sense of the market, is not easy. Few people can articulate how they do it. Good salespeople can't necessarily explain their techniques, but they know exactly when to close an order. Similarly, an effective marketing manager will say: "I just sense this is the way the market is."

Intuition is not magic. It is judgment based on experience. That type of sixth sense is acquired only by spending time with customers and in the marketplace. Marketers need to live and breathe the market. They need to talk to market participants on a continuing basis. Though ironic, it's true that in this age of electronic communications personal interaction is becoming more important than ever.

Some product marketing people spend their time putting together data sheets, writing application notes, conducting training sessions, and sending memos to the field. But memos don't a market make. In my mind, marketing people should be on the road half the time—meeting customers, talking to people, building relationships, and seeing where the next product should go.

Indeed, conversations with market participants often provide more insights than do a long list of statistics or a set of sophisticated theories. In marketing, experience is more valuable than logic. Students coming out of business schools think they're going to teach the world how to market. But experience is far more valuable.

In most aspects of Western life, logic is king. The Western approach to life encourages us to break things apart and analyze them. It embraces an engineer's view of the world: Things that can't be measured are irrelevant or illogical. This approach underrates the value of intuition. When you break things into little pieces, when you "statisticalize" them, you lose the intuitive feel that is so important. When you turn a perception into a statistic you rob the perception of its richness.

Relying on intuition scares people. But the most successful marketing people rely on intuition. John Sculley of Apple says that most of the important marketing decisions in his life have been based on intuition. Intuition should not be seen as negative. It is merely another form of knowledge. Intuition comes largely from experience, rather than from intellectual or analytic thinking. Information is gathered through the senses, sorted in the subconscious mind, and emerges as intuition. It might sound "soft," but intuition is just as valid as any other form of thinking.

When using an intuitive approach to marketing, it is important to look for patterns and trends and connections, not at raw numbers. Statistical studies might show that 10 percent of a company's customers are disappointed with its products. That doesn't tell the company whether that figure is growing or shrinking. Another study is needed to determine that. Nor does the figure tell the company how intensely disappointed the customers feel. Or whether the disappointed customers will soon become influential by vocalizing unfavorable comments about the company.

Focus groups can provide a certain amount of qualitative information. In these groups, potential customers directly express their ideas and opinions. It must be remembered that focus groups do not represent the actual buying situation. The participants know they are there to play a role. At too many companies the information from focus groups is quantified before it is distributed to decision-makers. This filtered information is not nearly as useful as direct information in understanding the attitudes and perceptions of customers. Key decision-makers must hear the opinions of customers directly.

Qualitative information can come from all types of sources. I have gained lots of useful information just by standing at the counter of an electronics or computer store. In the early days of calculators I saw a man take two calculators, one in each hand, to see which was heavier. He bought the heavier model. So I suggested to a client company that it put weights in its calculators. It did, and that seemed to help sales. The weights made the calculators feel more substantial. In a computer store I watched as a man told the salesperson that he wanted to buy a personal computer. The salesperson began asking very technical questions about operating systems, disk memory, RAM requirements, and so forth. The customer finally decided to hold off buying a computer until he learned more about them. The salesperson should have begun by identifying the customer's level of expertise.

You don't get qualitative information by asking 10,000 people to list their likes and dislikes. You get it by observing. When you go to enough retail stores you begin to understand the selling process. You see what customers are asking about. You see where customers hesitate. You begin to understand customer fears.

Many marketing managers get so wrapped up in their products that they're deaf to criticism and insensitive to the market. They spend all their time promoting the strengths of their product. They begin selling everybody and listening to nobody. They begin to think their view is the view of the entire marketplace. A marketing manager at Spectra-Physics, who recognized this problem, told me he got rid of one outside marketing consultant because "he began to believe us." It is critical to maintain a fresh and unbiased view of the marketplace. That view comes only by talking, and listening, to customers.

When you go out and ask questions you don't need to talk to a statistically significant sample. You just need to talk to the right people. I recently did some work for a fast-growing telecommunications company. Rather than conduct a survey of 500 people, I talked to the telecommunications managers at Coca-Cola, McGraw-Hill, and a dozen other major companies. I quickly discovered the key issues affecting telecommunications customers. Marketing managers should make regular trips to customer sites. It is probably one of the best market-research tools that exists.

We use these ideas when we perform external audits for our clients. The purpose of an external audit is to gather information and insights from the environment. That information is then used in developing a positioning strategy. The audit can act as a reality check on the perceptions that were expressed by company executives during the internal audit described earlier. Sometimes the external audit shows that the executives understand the market well. Other times it shows that company executives are out of touch with the realities of the market. In these cases a company's positioning plans must be drastically altered.

Key to a successful external audit is to use as interviewers people who have industry expertise. We use people who have a particular industry experience so that they can ask questions and probe beyond the superficial to get to the heart of the matter. During the audit we interview people from a number of different groups: existing customers, potential customers, distributors, industry experts, financial analysts, and perhaps key journalists. We typically talk to a dozen or so people in all. When conducting an audit for a semiconductor company, we might talk to five start-ups that will be buying semiconductors, five established semiconductor customers, and five industry analysts.

In these interviews, we don't look for specific facts and figures. Rather, we identify patterns, attitudes, and opinions that influence the thinking process. For example, we want to find out how open people are to the acceptance of new technology. Are they willing to try something different and new? We might ask questions such as:

- Of the products currently on the market, which do you like best? Why?

- Where do you think the market is headed? What are the most important trends?

- What do you think of XYZ technology? What are its advantages and disadvantages when compared with ABC technology?

- Which companies do you see as the rising stars of the industry? Why?

- When you buy this type of product, what factors influence your decision? How much are you influenced by cost? Ease of use? Added features?

- What do you see as the major limitations to growth in this market?

- Who do you see as the key opinion leaders in this industry?

- Does company XYZ provide the technology, support, and service you need?

Before getting its external audit, Convex, a start-up in the supercomputer business, thought the key feature distinguishing supercomputers from smaller machines was their ability to handle sixty-four bits of data at once. But the audit revealed that customers believed the ability to perform integrated vector processing was the key feature. As a result of the audit Convex also discovered a new market to target. It expected that people involved in seismic-activity exploration would be those most excited by low-cost super-computers, but it found that there was just as much excitement among people in the computer-aided design business.

External audits are particularly useful when they reveal trends and patterns for which the company was not even looking. To help position a new microprocessor for Intel, we talked to engineering managers at Hewlett-Packard, Xerox, TRW, and several other major companies. We asked about their expectations for the next generation of microprocessors. We found that the managers' ages affected the responses. The managers older than thirty-five were reluctant to try a new technology. They were more interested in quality assurance and documentation. The younger managers, on the other hand, wanted to experiment. We learned that Intel should present different messages to managers of different ages.

External audits should not be a one-shot deal. Companies must continuously monitor the environment to detect changes in mood and attitude. Everyone at a company, not just marketing managers, should be involved in the process. Engineers should meet regularly with customers, and so should top management people. Only through constant monitoring of the environment can companies stay on the right track.

Entrepreneurs usually start off on the right foot. They almost always have an intuitive feel for the markets to which they are selling. Top executives deal directly with their customers. They are constantly aware of what is happening in the market. This is a key reason for the success of so many small companies in fast-growing industries. These companies are more in tune with the market than their larger competitors.

But as they grow, corporations tend to lose this feel for the market. They begin to suffer from what I call "bigness mentality." Top executives forget their roots. Rather than rely on intuition, as they did in the early days, they start to manage by numbers. Qualitative information is replaced by statistical information. As staffs grow, they separate the top managers from the market. Managers begin to worry more about the efficiencies of mass production and less about the needs of the market. They no longer want to take risks. Their whole thinking process changes.

To continue with success, growing companies must continue to "think small." Managers must maintain their intuitive feel for market trends and attitudes. They should look at the numbers, but they shouldn't be ruled by them.

DECIDING ON A STRATEGY

Once a company understands the market environment it must decide on a strategy to get its products positioned within that environment. This is a fuzzy process with no firm rules to guide it. It is different in every case. Sometimes it evolves naturally. Other times it is the result of formal meetings and strategy sessions. In the best cases it is a combination of the two. This situation doesn't play well to managers who are looking for a systematic and clearly defined process. It would be nice if such a textbook approach worked consistently. However, the fact is that good business strategies come out of leadership, creativity, and an environment that encourages both.

Coming up with a massive document outlining the positioning strategy is not the way to begin the process. Marketing plans usually sit on shelves, gathering dust. People don't follow them day to day. Too often they are seen as substitutes for action. Rather, it is the ideas that are important. The company's key people must converge on a common positioning strategy, then implement it. Spending weeks or months writing down the ideas is simply a waste of time. Having regular work group meetings and review sessions is a much better way of insuring that the position strategy does not become static. It must become an active process. It should include adding to the team's knowledge, incrementally improving and probing ideas, and bouncing them off customers and members of the infrastructure. Positioning is an ever-changing process. It must constantly be worked on.

Positioning sessions with our clients have two primary goals. The first is to identify a position, and the second is to decide what actions are needed to achieve the position. As a result of the sessions, many things can change. Positioning it is not just a matter of coming up with a new slogan. By reevaluating its positioning, a company might change its market direction, its target customers, its distribution strategy. It might even change the products themselves.

We run positioning sessions after completing the internal and external audits, so we already have a good feel for the internal dynamics of the company and the attitudes of the marketplace. We walk into the meetings armed with direct, qualitative information from the marketplace. We have talked to customers and the various members of the industry and market infrastructure. The challenge is to use this information to help the company target its product and thus develop a unique position in the marketplace. We must decide how to differentiate the product from its competitors, how to distribute and promote it, and how to gain credibility for the product in the marketplace.

Typically we meet with six to ten people from a company. The people represent different groups and different experiences. Some people are from the sales organization because they continually interact with the customers, getting information direct from the market. But their views are somewhat limited: Selling is a one-to-one process, and marketing must consider the broader picture. The meetings also generally include some people from the technical side of the company. They, too, have a somewhat narrow perspective, but they can mark a discussion's boundaries. They don't let us go too far with our analogies.

Each person can relate individual experiences. One might say: "When I was at DEC, such and such happened." Another will add: "When I was at Hewlett-Packard, we did it this way." None of these past models will represent a perfect positioning strategy. But some mixture of these experiences should lead to the right positioning approach.

Interaction at the sessions usually is not systematic. It involves more brainstorming, or free association. The goal is to maximize creativity. Everyone tosses out ideas, then others modify the ideas and add to them. In running a session, I try to listen to a variety of experiences and views, without allowing any one of them to dominate my thinking. I make sure not to reach conclusions too quickly. I don't want to get locked in place and lose flexibility. Once a conclusion is drawn, people begin to argue it, rather than look for new insights. The phenomenologist philosopher Husserl advised: "Bracket your prejudices." I try to follow that advice. I keep prejudices out so reality can penetrate.

Ideally, the sessions break neatly into three stages: input, analysis, and synthesis. Each stage can be an hour, or a half or full day, depending on the complexity of the problem and the experience of the team.

In the first stage we listen to presentation and company analysis. We take notes and absorb information. We look for patterns and connections. These don't always come quickly. We wait, wait, wait. We wonder whether there's a different way, or a better way, of doing something. We encourage off-the-wall ideas. We look for things that the other people know intuitively but haven't been able to articulate.

We look for all types of relationships. How does the product relate to past and future products? How will salespeople and customers relate to one another? How does the operating system relate to the applications program? How does the telecommunications manager relate to the MIS manager? How does the company relate to its suppliers? We are always looking for ways in which the company can use these relationships to its advantage.

In the second stage we write some of the ideas on the board. We list obstacles. Competitors. Environmental factors. We put all types of things on the board and try to relate one thing to another. We bring up examples from other companies, and add some ideas about industry structure. Much of our information is derived from talking to customers or members of the infrastructure. That way we can act as surrogates for the customer.

Sometimes we use the idea of "positioning denominators." We make a chart listing strengths and weaknesses in the three positioning categories: product, market, and corporate. In the product category we compare the product to its competitors in terms of power, speed, and compatibility. In the market category we compare distribution, sales forces, and customers. In the corporate category we compare financial resources, reputation, and management image.

The third stage is devoted to synthesis. We don't use numbers or graphs, we just manipulate ideas. Working together, we try to integrate all the ideas brought up by the salespeople, technical people, and others. We link together the strengths among the positioning denominators, and try to turn them into a coherent plan.

We battle back and forth until we hit on the right positioning plan. It usually comes in an "Aha!" experience. All of a sudden everything makes sense. Everyone agrees: "That's it! That's right!" Out of the murky mess of information a clear vision of the future has emerged. I've gone through the process hundreds of times, and we've reached that type of conclusion in all but a couple of cases. However, the conclusions are never absolute or final. A review agenda is established so that the process of change and improvement can take place.

One of the failures involved Imagic, a now-defunct company that produced video-game software. We met with Imagic managers for a total of six hours, but we could not find any perceptible differences between their strategy and the strategy of their industry's leader. None. They simply said: "We're going to make games and take market share away from the leader." Their goal was to get a lot of sales and to go public fast. There was no real competitive differentiation between the two companies. If one had a "Turtle Walk" game, the other would develop a "Rabbit Walk" game. The differences were all superficial. Imagic had no solid basis for differentiation of itself or its products. The company eventually went out of business.

In most cases, though, the combination of internal audit, external audit, and positioning sessions leads to a clear positioning statement. The result of meetings with Convex, the supercomputer company mentioned earlier, provides a good example. The company has a strong technology group, led by Steve Wallach, one of the star engineers portrayed in the book *Soul of a New Machine* by Tracy Kidder. When they came to us, however, the company founders had not built a consensus of and had not clearly articulated their desired position in the market.

Convex initially planned to position its computer as a superminicomputer. That product category was established in the late 1970s when Digital Equipment Corporation introduced a line of computers, known as VAX computers, that were more powerful than traditional minicomputers. Within a few years superminis had become very popular, particularly for use in scientific applications. Convex was designing a machine that would run the same software as a VAX, but would be even more powerful.

There was one big problem: The superminicomputer field was already quite crowded. There were more than fifty companies selling superminicomputers. Convex's machine was probably better than the rest, but it would be difficult to make the machine stand out in that market.

So we began our external audit and looked for ways to differentiate the Convex computer. We found a tremendous demand for superminis like the VAX, but also growing customer dissatisfaction. After many years as the workhorse for scientific applications, the VAX was starting to show its age. Scientific problems, such as the design of very-large-scale integrated circuits, were becoming more complex, and VAXs could no longer handle the job very well. Many users wanted the high speed and special capabilities of a supercomputer. But existing supercomputers, sold by Cray and a few other companies, had drawbacks. They were too expensive for most applications. And they couldn't run nearly as many different types of software programs as the VAX could.

At the positioning session, we quickly recognized that there was a huge gap in the market. On one side was the VAX: lots of software, prices ranging from $500,000 to $2 million, but insufficient power for many new scientific applications. More than 50,000 of them had been sold. On the other side were Cray-like supercomputers: lots of power, but not much software and prices as high as $5 million to $10 million. Fewer than 150 of these computers had been sold.

We saw an opportunity to open an entirely new market segment between the VAX and the supercomputers. There was a huge gap in price and performance, and the Convex computer could fit right in it. This computer was twenty times faster than a VAX, but only one-quarter the price of a supercomputer. It could run all the VAX software, but it also could provide almost as much power as a supercomputer. In the future, the computer's performance could be incrementally expanded into a more powerful supercomputer.

The question then became: Should we position the computer as a super-VAX or as a baby supercomputer? We quickly decided that it made more sense to position the computer as a baby supercomputer. Rather than compete directly against fifty suppliers of VAX-like machines, including powerful DEC, Convex would be positioned in the supercomputer industry, a market segment with only three or four manufacturers.

The technology had not changed at all, but the marketing plan was now totally different. The company has begun to think of itself differently. It is in a Cray-like business, not a DEC-like business. That means using different types of pricing and different types of marketing. Convex managers now pay attention to all issues relating to supercomputers. For example, the press have become quite interested in the Japanese efforts to leapfrog past the United States in supercomputer technology. Now Convex can share in that limelight. Convex's president was invited to participate in a roundtable discussion on the Japanese challenge, giving the company unexpected visibility. Suddenly, Convex is not just an innovative company, it is an American asset in an important international competition. It's amazing what a little positioning strategy will do.

The positioning influenced the development of the next generation of Convex computers. The capabilities of the new products challenged Cray Computers so effectively that Cray had to revise its product strategy to compete with Convex. Cray's repositioning reinforced the Convex position. The company enjoys a strong financial performance, and is now well recognized as a leader in the supercomputer field.

Start-ups are not the only companies that need to formulate positioning strategies. In dynamic markets, companies must constantly reevaluate their positioning plans. A positioning plan that seems to make sense one month can be thrown into disarray the following month as new products come to the market and customer attitudes change.

Intel faced this problem with its highly successful 8086 microprocessor. After its introduction in 1978, the 8086 quickly became an industry star. It gained a dominant market share in the market for 16-bit microprocessors—that is, microprocessors that can handle sixteen bits of data at one time.

By 1981, however, storm clouds were gathering. Motorola had introduced a competitive chip in late 1979, and the Motorola chip, the 68000, was beginning to attract attention in the industry. Motorola had followed a classic "second-company-in" strategy. It had overcome some of the weaknesses of Intel's 8086 in its own product, then tried to grab onto the market momentum the 8086 had created.

Intel's marketing staff had been slow in identifying the new trend. They were sitting in Santa Clara looking at industry growth charts and design wins, and they didn't see any big problems. But Intel salespeople began to see something different in the field. The 8086 was still selling well, but customers clearly were intrigued with the Motorola chip. The salespeople had to work harder to sell the Intel 8086. They sensed that momentum was shifting from the 8086 to the 68000.

Clearly, Intel needed to do some repositioning of the 8086. Its problem was more than a chip versus chip battle. Entire product lines were at stake. If a microprocessor sells well, it emerges as a standard and all the other chips in the product family are carried on its commercial coattails. The battle between the 8086 and the 68000 was really a battle between Intel and Motorola for industry leadership.

To get a better feel for customer attitudes, we did a quick external audit of the market. Intel president Andy Grove called together a special team to address the problem. The team consisted of six Intel managers, from marketing, engineering, and sales, and myself. The group was lead by Bill Davidow, then head of Intel's microprocessor operations and now general partner of Mohr-Davidow, a high-tech venture-capital firm. Casey Powell, another member of the group, was a regional sales manager and the person who initiated management action by writing a memo to Grove pointing out the problems salespeople were facing getting new design wins against Motorola. (Powell went on to found Sequent Computer, a highly successful computer company.) We met for three days straight, from a Wednesday through a Friday.

Our repositioning project went by the code name CRUSH. Jim Lally, now a partner at Kleiner Perkins Caufield & Byers, came up with the name. He wanted to give the project an emotional and competitive punch. Our mission was straightforward: Identify why the Intel 8086 was beginning to slip in its competition against the Motorola 68000, then implement a strategy designed to respond and recover.

We began by dividing the market into different types of customers. We segmented the customers not according to market size or location, but according to their thinking processes and attitudes. We decided that some customers were hardware oriented. They cared most about raw performance factors, like speed and power. Other customers were software oriented. They had much different priorities. They wanted a microprocessor with a straight-forward architecture so it would be easy to develop software for it.

The external audit, as well as information obtained from Intel sales-people, indicated the Intel 8086 was holding its own among hardware-oriented companies. But the Motorola 68000 was quickly gaining market share among software-oriented companies. Software developers felt more comfortable with the Motorola chip. They felt it provided more support and flexibility for developing new applications. We didn't have specific statistics to support these findings, but our qualitative information left little question in our minds.

The challenge, then, was for Intel to reposition itself among software-oriented companies. We decided on several ways to do that. One way was to focus more on the breadth and depth of Intel's product line. As long as customers continued to focus on chip versus chip, 8086 versus 68000, Intel would have troubles. In a horse race among chips, Motorola would win among certain customers. But if customers looked at overall solutions and future directions, Intel would have advantages. Intel's 8086 could be combined with its 8087, for example, to provide the best solution for use in scientific applications. It could be combined with a different Intel chip for use in a communications application.

By focusing on the whole product line we also hoped to get customers thinking about the future. We wanted people to worry about the consequences of committing themselves to Motorola. We wanted to play on the customers' fears. Sure, Motorola had a hit product, but could the company support it with other chips and future enhancements? The 68000 had almost no software to use with it, no peripheral chips, no development system. And Motorola had not explained its future plans. By committing to the 68000 architecture, might customers get stuck in the future?

Intel, by contrast, already had a full family of microprocessor products. It was a safe bet for the future. To reinforce this point, we planned to show customers Intel's plans for future-generation microprocessors, both up and down the product line. The message would be clear: Intel had a well-developed plan for the future. With Motorola, the future was murky.

We also identified another Intel advantage. Intel's top executives—Bob Noyce, Gordon Moore, and Andy Grove—were perceived, accurately, as pioneers and innovators in semiconductor technology. Their credibility was high. If they talked directly to major customers their message would carry a great deal of weight. We planned to have the top three and other leading technical people make presentations at small seminars. These seminars would require a lot of valuable time, but they would make a much stronger impression on customers than would advertisements and articles. Intel would come across as having a great deal of technological depth, just by virtue of the people who gave the seminars.

Intel wasted no time putting CRUSH into action. Our group finished its three-day positioning session on Friday. The following Tuesday the group presented its findings and requests for budgets to the executive staff of Intel. On Wednesday we assembled more than 100 Intel managers from all over the world in order to explain the project to them. Each was assigned a specific task—a software task, a technical task, a documentation task, an advertising task. It took Intel less than seven days to develop a new positioning strategy and put it into place. This ability to respond quickly is an important corporate asset. Some time later, when I told a former Motorola executive that it took only seven days to develop the CRUSH program, he told me that Motorola could not have even organized a meeting in seven days.

Over the next three months Intel executives gave presentations to more than thirty major customers. In the following quarter Intel gave nearly fifty half-day seminars to potential customers. Gradually, the market's momentum shifted away from the 68000 and back toward the 8086. Motorola had a strong technical product, and its sales continued to grow. But Intel had won the positioning battle. Its 8086 remained the leading 16-bit microprocessor in the industry and has sustained that leadership for over a decade.

The CRUSH program became an ongoing process at Intel. Much of what was learned, such as the need to involve the whole organization, real-time competitive analysis, how to integrate resources in order to solve marketing problems, the value of bouncing ideas off key customers, and the need for a continuing review of marketing programs, was incorporated into the marketing process.

There is no special magic in what Intel did. This same approach can work for other companies and other industries. Many companies, over time, lose their capacity to act, losing both the value of the market momentum and the timeliness of response. Companies that plan qualitatively and react swiftly always will be a step ahead of the competition in the battle for strong market positions.

> Wooden-headedness, the source of self-deception....It consists of assessing a situation in terms of preconceived fixed notions while ignoring or rejecting any contrary signs.
>
> Barbara Tuchman
> *The March of Folly*

| # Things that Go Bump in the Night: The Ten Biggest Competitors

INTANGIBLE COMPETITORS

Ask marketing managers to name their primary competitors and they'll rattle off the names of a few other companies in the industry. Marketing managers in the personal computer industry worry about competition from IBM, Apple, and Microsoft. Those in the semiconductor industry worry about Nippon Electric (NEC), Fujitsu, and Texas Instruments.

These worries are, to a large extent, misplaced. Certainly, all those mentioned are tough competitors. But they aren't the toughest competition. They aren't the *real* competition.

The real competition comes from what I call "intangible competitors." These competitors involve ways of thinking and ways of looking at the world. They are obstacles that get in the way of success. When marketing managers resist change, they are facing an intangible competitor. When entrepreneurs begin thinking in the bureaucratic style of "large-corporation man," they are up against an intangible competitor.

Not handling these intangible competitors skillfully is the primary reason marketing fails. If companies can deal with these competitors, they are better equipped to succeed. I have identified ten intangible competitors that all companies confront, regardless of what industry they are in. They are:

1. Change

2. Resistance to change

3. Educated customers

4. The customer's mind

5. The commodity mentality

6. The bigness mentality

7. Broken chains

8. The product concept

9. Things that go bump in the night

10. Yourself

Competitor 1: Change

Our society is in a perpetual state of change. Everything is changing. Companies change. One day the newspapers carry a story about a computer company hitting $100 million in sales. A few weeks later they carry a story about the same company going bankrupt.

Industries change. Deregulation has had a major impact on the structure of the telecommunications and airlines industries. It will completely alter the financial and banking industries in the next decade. Global competition has changed the way companies do day-to-day business, compete, and get financing. One decade ago the names Microsoft, IBM, Compaq, and Sun had no meaning in the small-computer business. Today they are the major influences in the industry's technology and markets. The software industry has undergone an even bigger change. Ten years ago the industry included a few hundred companies. Today there are thousands.

Products change. Today it seems that every product is becoming "smart." Microwave ovens have microprocessors in them. Telephones have microprocessors in them. Even toys have microprocessors in them. With these microprocessors tucked inside, familiar products take on new traits and perform new tasks. Computers themselves are changing too. Today we have computers in all shapes and sizes—personal computers, hand-held computers, portable computers. A decade ago there were few more than 100,000 computers installed around the world. Today 50,000 computers are bought every day.

Distribution channels change. A decade ago no one believed that a computer could be sold through a retail store. Today retail stores sell hundreds of thousands of computers every month. Certainly no one would have believed that a large business could be built based on selling computers by direct response. Dell Computer showed it could be done very well. The value-added reseller and the systems-integration business have also become channels not foreseen a decade ago.

Issues change. Industry issues keep changing as new technologies transform the way we approach our work. Ethics has become a concern of many companies, particularly those in the biotechnology industry. Ethicists, whose main source of employment has been teaching at universities, are now in demand throughout the business world. The shift from producer-driven markets to consumer-driven markets has occurred rapidly in the past decade, creating a host of new issues. The idea of "customer satisfaction," championed in the 1980s by people such as Tom Peters and J. D. Power, has become a major issue for corporations trying to figure out what it means and how to implement suitable programs. In the computer industry, the CISC battles of the 80s are giving way to the RISC battles of the 90s. (Both CISC and RISC are forms of computer architecture, with RISC being the newest generation. With the change to RISC, it is expected that the industry will see new performance levels in computing, new software, and new players in the industry.)

These changes constitute a major competitive force. They have a deep influence on the growth and direction of every company. Companies blind to change are doomed to failure. Not keeping pace with change can topple even dominant companies. Companies simply can't afford to stay in the same place.

Business history is full of examples of companies that didn't recognize change in the market, and paid a heavy price as a result. For years, the U.S. auto companies ignored the growing demand for small cars. Japanese companies were attuned to the changing market, though, and they quickly stole market share from their American rivals.

The story is similar in the semiconductor industry. In the early 1960s, Fairchild, Philco, and General Electric were dominant forces in the industry. None of them recognized the growing importance of integrated circuits, however, and not one of them is a major factor in the industry today. And the process goes on. Ten years ago semiconductor companies felt pretty secure. They believed the capital intensity of semiconductor manufacturing would prevent new companies from entering the business. But the technology changes allowed the unbundling of the design of the chips from the processing. This change in the technology created a whole new cadre of "fab-less" semiconductor companies (companies that design chips and have the processing done at silicon foundries). Those companies that didn't acknowledge this change have been left behind.

The computer industry offers another example. All the major computer companies ignored personal computers in the 1970s. Small start-ups began selling personal computers in 1976, but big companies such as Digital Equipment simply didn't anticipate a market change. But this market revolution took place right under industry giants' noses. Many large computer companies, including Burroughs, Honeywell, Wang, Data General, Hewlett-Packard, and NCR, were put in a position of following rather than setting the standards of the new industry. Several have never fully recovered.

In his autobiography, *My Life and Work,* Henry Ford said, "I saw great businesses become but a ghost of a name because someone thought they could be managed just as they were always managed, and though the management may have been most excellent in its day, its excellence consisted in its alertness to its day, and not its slavish following of its yesterdays." Change has become a part of our lives, with one thing inexorably replacing another. We destroy the old and create the new. In all industries, change is a tough competitor. What can managers—all managers, those in sales, manufacturing, engineering, and marketing—do to cope with this competitor? Two things.

First, managers must constantly question their assumptions. They must ask questions such as: "What am I assuming about the market?" "What am I assuming about the competition?" "What things must happen to make my assumptions valid?" "Under what conditions are my assumptions no longer valid?"

Second, managers must keep their ears to the ground. They must sense change as it is occurring. They must monitor the market, live with it, work with it. Oftentimes changes don't show up in the numbers and statistics until it is too late to respond to them effectively. Marketeers must develop an intuitive sense of the market. They must work with customers and listen to them. They must meet with dealers and listen to them. And they must *really* listen. That is the only way they will spot changes in time to adjust.

Competitor 2: Resistance to Change

Sometimes companies recognize that change is occurring in the marketplace, but they still don't react. For these companies the competitor is resistance to change. Resisting change can be just as damaging as being oblivious to change. In either case, the company can get left in the technological dust.

Examples of resistance to change abound. Consider the case of Gary Boone. In 1972, as a young engineer at Texas Instruments, Boone came up with the idea for a full computer on a chip, later to be called the microprocessor. Boone got a patent on his invention, but he had trouble getting his colleagues interested in his work. He went around TI trying to sell the concept, but he was shot down everywhere. Other people looked at him as a young guy with a crazy idea.

Finally, Boone made enough noise to get a meeting with TI's top guru on computers. Boone went into the office, sat in front of the expert, and explained his idea for a computer on a chip. The expert looked at him with a condescending smile. "Young man," he said, "don't you realize that computers are getting bigger, not smaller?"

There are similar stories involving personal computers. Steve Jobs and Steve Wozniak tried to sell the idea of personal computers to their bosses at Atari and Hewlett-Packard. But their bosses weren't interested. So Jobs and Wozniak started Apple Computer. Intel also had a chance to get in on personal computers early. Several Intel marketing pros went to visit one of the early designers of personal computers sometime in the mid-1970s. They came back and reported: "Bunch of hobbyists. It will never be much of a market."

Such resistance to change can destroy companies. Take a look at the American industries involving the following product lines: autos, steel, consumer electronics, calculators, machine tools, and textiles. In the mid-1960s imports accounted for less than 10 percent of sales in the U.S. market in each of these industries. But American companies in these industries became resistant to change, while their foreign competitors did not. The result: In 1981 the United States imported 26% of its cars, 17% of its steel, 60% of its consumer electronics (television, stereos, videocassette recorders), 41% of its calculators, 53% of its machine tools, and 35% of its textiles.

What makes companies resistant to change? Sometimes bureaucracy is to blame; sometimes it's just that people are scared and intimidated by new things. People tend to get wedded to ideas. They look toward the past, rather than toward the future. When employees move to a new company or a new project, they bring their histories with them. This experience can be useful, but it also can cause problems. Marketing people often say "This is the way we did it at my old company." This is helpful sometimes, but every once in a while they should say: "Let's experiment and try something new."

The resistant-to-change demon rarely haunts young entrepreneurial companies. Entrepreneurs thrive on innovation and change. They are always willing to experiment with new ideas and new technologies. Resistance to change is anathema to entrepreneurs.

As entrepreneurial companies grow, however, they become more resistant to change. They begin to think more about high-volume production. They develop inflexible systems, processes, and ways of doing business that commit them to doing things in a repetitive, predictable way. That behavior locks them into offering certain products and technologies. They begin to ask questions such as "How do I keep my factories going?" and "How do I keep selling at this rate every month?"

In short, production becomes the central focus of the company. The company begins to worry more about organization and less about serving the needs of the customers. As a result, the company takes on the personality of a large company and becomes less likely to develop innovative new products. Small companies grab the lead in innovation.

The scenario is repeated time and again. The semiconductor memory business provides one example. Intel developed the first semiconductor memory chip, the 1K RAM. It clearly established itself as the leader in this new product category. But when Intel began working on the next generation of memory chips, the 4K RAM, it lost its innovative edge. Intel was committed to the development approach it used with its money-making 1K RAM, but other approaches were better suited to the new generation. A small company called Mostek developed a more innovative 4K RAM, and it emerged as the new leader. And because the two companies did not collaborate on a single standard and work together, the Japanese became the eventual victors.

Mass production inhibits change and innovation. It is based on stability and predictability. When a company moves into mass production it becomes resistant to change. It wants to build up economies of scale. Innovation can disrupt that effort.

Growth companies face a difficult challenge. They must figure out how to move toward high-volume production without losing the innovative spirit that made them successful in the first place. They must continue to see innovation and change as allies, not as competitors.

Corporate politics plays a major role in inhibiting change, particularly in American corporations. Power in the executive ranks seems to be more of a driving force than power in the marketplace. Recently appointed president of Apple Computer, Mike Spindler likes to ask people whether they can name the presidents of several very large Japanese companies. Most people can't. But they do recognize the companies and the industry leadership positions these companies hold. American company managements are more concerned with the appearance of position. Research is done and facts are garnered to support positions rather than to identify weaknesses. Cheerleading is a regular ritual. Dave Power of J. D. Power told me that Japanese managements don't care for studies that show a simple ranking of car makers based on customer satisfaction. The Japanese managers want to know what they are doing wrong so they can fix things. They want the research to involve asking customers to identify problems, concerns, likes, and dislikes. Things the company can improve on. American car makers like the simple ranking. Even when an American car is far down the customer-satisfaction list, the manufacturer will find a way to brag about it.

Competitor 3: Educated Customers

An uninformed customer is easily satisfied. But there aren't many uninformed customers around these days. Customers today have access to more product information than ever before, and they study it carefully. With technology products, customers are becoming more technology literate.

Customer technology literacy presents a challenge to manufacturers. Customers are no longer pushovers. They want to understand more about the products they buy. They are skeptical and critical, and more often dissatisfied. Manufacturers must meet a higher level of expectations.

Consider the amount of computer information reaching the public these days. A few years ago there were a handful of computer magazines. Now there are hundreds. A few years ago *Time* and *Newsweek* magazines hardly ever wrote about computers. Now they both have computer editors. A few years ago television news never ran stories about computers. But for the introduction of the Macintosh, all three major networks ran stories, as did more than twenty individual stations. Information on high technology has become as integrated into the news as information about the auto industry.

As the quantity of coverage has increased, the quality of coverage has improved. Journalists themselves are becoming more technology literate. Until a short time ago, computer companies could use journalists to spread just about any message they wished. The journalists didn't know enough about technology to critically evaluate computer companies and their products. That has changed. Many journalists use personal computers and are quite knowledgeable about them. When a company introduces a new computer today, journalists want to evaluate the computer themselves. They won't take the company's word about what the machine can and can't do. In effect, the journalist becomes an evaluator for the public.

Hundreds of on-line data bases are available to everyone. We live in the information age and we are often smothered by an excess of information. However, over the next decade information will become increasingly specific and customer programmable. The customer will become even more powerful because the computer will become more useful as an information-sorting and decision-making tool.

To succeed, companies must turn customers' increasing knowledge of their products from an obstacle into an asset. Rather than battle against a skeptical, critical, and uninformed public, companies should learn from it. They should elicit feedback from customers, then adjust their products and strategies to meet the market needs. An educated customer can only make a business more competitive.

Some consumer-goods companies are already quite successful at using customer dissatisfaction to their advantage. According to a *Wall Street Journal* article, Procter & Gamble phones or visits 1.5 million people each year to ask about P&G products. P&G researchers ask hundreds of detailed questions to find out why customers are dissatisfied and what actions P&G should take to improve its products.

The same article quoted from a study by the U.S. Office for Consumer Affairs: "Many managers view complaints as a nuisance that wastes valuable corporate resources. However, the survey data suggest that complaints may instead be a valuable marketing asset. Responsive companies were rewarded by the greatest degree of brand loyalty."

Technology-based companies should learn a lesson from this. As customers become more knowledgeable—and more critical—about technological products, companies must become more sensitive to customer needs. The philosopher John Stuart Mill once said: "Better to be Aristotle dissatisfied than a fool fully satisfied." Customers of technological products are taking Mill's advice, and companies must adjust.

Competitor 4: The Customer's Mind

People in technology-based businesses tend to think decision making is a simple and rational process. They are wrong. Indeed, when a customer considers buying a product, the decision-making process is neither simple nor rational. All types of fears, doubts, and other psychological factors come into play. Information comes in many disguises.

Winning over the customer's mind is the central challenge of marketing. The customer's mind can be seen either as a competitor or as a competitive tool. Sometimes the customer's mind can act as an obstacle to success. But if companies can understand the customer's mind, they can use psychological factors to their advantage.

All types of things influence the customer's mind. Indeed, the battle for sales is largely a psychological battle. As I explained before, decisions are based largely on intangible factors such as quality, image, support, and leadership. In *Future Shock,* Alvin Toffler describes the psychological battle this way: "For even when they are otherwise identical, there are likely to be marked psychological differences between one product and another. Advertisers strive to stamp each product with its own distinct image. These images are functional. The need is psychological, however, rather than utilitarian in the ordinary sense. Thus, we find that the term 'quality' increasingly refers to the ambiance, the status associations—in effect, the psychological connotations of the product."

Customer attitudes toward a product are not developed by a single event or a single advertisement. Rather, customer attitudes develop gradually. They are constantly changing and evolving throughout the decision-making process—and continue to evolve after the decision is made. The "product image" is simply the accumulation of all these attitudes.

The customer's mind can be influenced at every step during the decision-making process. First the people become aware of the existence of the product. Then they recognize the need for the product. At that point they will try to find out more about the product. They might talk to people who already have used the product, or read reviews written by experts. They might use the product on a trial basis. At each stage their attitudes are modified and reformed. After the purchase, customers' attitudes continue to evolve as they use the product. Customers expect a certain level of product support and product performance. If support and performance fall short of expectations, customer attitudes toward the product and the company will turn negative.

Throughout the entire process, "psychological bogeymen" affect the customer's mind. These bogeymen include all types of doubts and fears that surround the product, making the customer uneasy about making the purchase. Customers might worry about such things as:

- Is the company going to be around for a long time?

- Am I going to be able to get product support after the purchase?

- Will the manufacturer be able to supply future generations of products?

- Will I be technologically behind if I buy this company's products now, rather than wait for its competitor's upcoming product?

In winning the battle for the customer's mind, companies must fight against these psychological bogeymen. They must provide comfort factors that put the customer's mind at ease. For marketing complex technical products, these comfort factors are particularly important. A company must convince customers that it is financially and technically strong enough to meet all of the customers' future needs.

At the same time, companies can try to influence customer attitudes toward competitors' products. With its FUD strategy mentioned earlier, IBM works both strategies. It surrounds its own products with comfort factors and introduces psychological bogeymen to its competitors' products. Clearly, the strategy has been quite successful.

To succeed in a market, companies also must work to understand the customer's mind. It is not enough to know what competitive products are on the market and who is using them. Marketeers must also understand the psychological bogeymen and comfort factors that influence the customer's mind, then use these psychological factors to their advantage.

Competitor 5: The Commodity Mentality

What is good for manufacturing is not always good for marketing. For efficient, low-cost manufacturing nothing beats commodities. By churning out the same commodity product time after time, manufacturers can work all of the kinks out of the production process. As volume increases, manufacturers move down the so-called learning curve, and their costs drop lower and lower.

But a marketing strategy that depends on a commodity mentality can be deadly. Customers usually prefer custom-made, "just-for-me" products. They want their needs satisfied exactly. We are in an age of diversity, and people want to feel as if they are getting something special.

Companies that view their products as commodities will have an increasingly difficult time competing, especially in evolving markets. Companies that sell commodity products can attract customers only by keeping prices low. Competition generally degenerates into a struggle for price leadership, and no one ends up making much money.

How can companies get out of this commodity trap? Meshing the differing needs of manufacturing and marketing isn't always easy, but it can be done. The trick is to view products as more than physical entities. Even if a company manufactures commodity-like products, it can differentiate the products through the service and support it offers, or by target marketing.

It can leave its commodity mentality in the factory, and bring a mentality of diversity to the marketplace.

To move away from the commodity mentality, companies must view their products as problem solvers, and then sell the products on that basis. Service adds another dimension that provides commodity businesses with a differentiation. Dell Computer sells IBM clone computers. But by also offering a twenty-four-hour 800 number for support, service guarantees, information via fax, and other services, it has achieved a distinguished position in a commodity business.

In his *Harvard Business Review* article "Marketing Success Through the Differentiation of Anything," Theodore Levitt describes the approach this way: To the potential buyer, a product is a complex cluster of value satisfactions. The generic thing is not itself the product .... A customer attaches value to a product in proportion to its perceived ability to help solve his problem or meet his needs.

An automobile, for example, is not just four wheels and an engine. It is a product that fulfills customer needs, psychological and otherwise. Manufacturers can differentiate their automobiles according to the needs they fulfill. One can be positioned as a status product, another as a performance product, even if the products themselves are quite similar. If automobiles were marketed solely on the basis of their specifications (the number of cylinders, the size of the engine, and others), customers would perceive them all as being very much alike. Indeed, specsmanship marketing is a sure sign of a commodity mentality.

The personal computer provides another example. Everyone views the personal computer in a different light. Many managers see it as a productivity tool that provides increased freedom to information users. Some MIS managers see the personal computer as a device that causes information and other resources to be used inefficiently within large organizations. The product is the same, but the perceptions of it are very different.

The perception of personal computers also changes with time. At first the Apple II was seen as a hobbyist computer. Then as a small-business computer. Then as a vertical-market computer—a computer able to serve many different, specialized applications. But the Apple II itself remained largely the same. However, the marketplace has changed, and so has Apple's marketing strategy. Apple has manufactured the Apple II like a commodity. But in its marketing, Apple made the Apple II special to every customer. It stayed away from a commodity mentality.

Competitor 6: The Bigness Mentality

Edward Schumacher, the economist, was certainly right when he coined the phrase "small is beautiful." Just consider the following statistics:

- More than half of the innovations in the United States in the last twenty years have come from companies with fewer than 200 employees.

- A study by Massachusetts Institute of Technology professor David Birch showed that companies with fewer than twenty employees created 60% of all new jobs, and companies with fewer than 500 employees created 86% of all new jobs.

- Of the 9 million jobs created between 1966 and 1977, 6 million were created by small businesses, 3 million by government, and zero by Fortune 1000 companies.

- Small companies are more efficient with R&D. A study by the National Science Foundation showed that small companies (those with fewer than 1,000 employees) produced four times as many innovations per R&D dollar as medium-sized companies (those with 1,000 to 10,000 employees) and twenty-four times as many innovations as large companies (those with more than 10,000 employees).

Indeed, study after study show that small companies are more innovative and productive than larger companies. Unfortunately, as small companies grow and become large companies, most of them run into the same problems as other big companies. They become less creative and less dynamic. They begin to suffer from what I call bigness mentality.

A major element of bigness mentality is an aversion to risk. Small companies can't afford to take the safe path. They could not compete with established companies on that basis. They must come up with new ideas, experiment with new approaches, try new things. They must innovate or they will not survive.

As companies grow they become more reluctant to take risks. If a company decides to go public, as most do, it is evaluated by the financial community on a quarterly basis. If financial results slip during one quarter, the stock price could plummet and the company could have trouble raising new funds. So public companies must play it safe. They can't afford to take short-term risks, even if they might pay off with long-term benefits. Wall Street thinks short term, not long term.

Corporate bureaucracy also reduces risk-taking and innovation. As small companies grow, they restructure themselves to look and act like big companies. Decisions are made by committees, not by individuals. As a result, decisions tend to be compromises, not bold new approaches. People begin to worry more about avoiding mistakes than creating new ideas.

Take advertising decisions. Advertisements developed by small companies tend to be much better than those developed by big companies. They are more creative, more aggressive, more interesting, more attention grabbing. Why? Big companies usually have large advertising departments that make decisions by committee. All decisions must be supported by extensive research. People are not willing to stick their necks out.

As a growing company adds new committees and new levels of bureaucracy to its ranks, it is slower to notice new opportunities in the market and slower to respond to changes in the market. Its corporate reaction time shoots up. Earlier I discussed Intel's quick reaction to the challenge from Motorola's 68000 microprocessor. Within seven days Intel designed a new strategy, presented the plan to 150 managers from around the world, and began to put the plan into action. At most large companies it probably would have taken seven days just to arrange the initial meetings.

How can companies avoid succumbing to the bigness mentality? One way is to maintain small, entrepreneurial project groups within the company. IBM took this approach when developing its personal computer. The company created an independent group in Boca Raton, Florida, and gave the group an unusual degree of freedom. In doing so, IBM acted in an un-IBM way. Although IBM was still a big company, it was thinking like a small one. It broke its own rules, and took some risks. The risks certainly paid off.

Companies also should avoid compartmentalization in the corporate organization. Many growing companies break various functional groups into different divisions, then make it difficult for those divisions to interact. In small companies people in the engineering, marketing, and sales departments talk regularly and exchange ideas. They act more as a team. "Communication does not work well if the group is very large," says Peter Drucker in his book *The New Realities*. "It requires constant reaffirmation. It requires the ability to interpret. It requires a community." This interaction is vital to developing creativity and innovation, but it is usually missing in large companies.

In his book *The Next American Frontier*, Harvard professor Robert Reich argues that large companies must develop new forms of organization that allow greater interaction among different groups. He writes that "precision manufacturing and custom-tailoring, and technology-driven products have a great deal in common. They all depend on the sophisticated skills of their employees, skills that are often developed within teams. And they all require that traditionally separate business functions (design, engineering, marketing, and sales) be merged into a highly integrated system that can respond quickly to new opportunity. In short, they are premised on flexible systems of production." Only with this type of flexibility can companies avoid the bigness mentality and maintain their creativity and productivity as they grow.

Competitor 7: Broken Chains

The business world is full of chains and connections. Processes and products are linked to one another in a great chain that ultimately connects companies and customers. No problem or business decision is isolated or self-contained. Companies get into trouble when they think about one link at a time, focusing on advertising or channels or manufacturing, without recognizing that all these functions are interrelated. By ignoring the linkages companies end up with a broken chain—and a failed product.

To get more specific, consider the most important chain: the product–customer chain. This chain connects everything in the product–development and marketing processes. It starts with the design and planning of the product. Other links include product development, manufacturing, marketing, sales, distribution, product support, and service. The final link in the chain is the customer. Here, the marketplace is a system that must be designed to meet the needs of a specific group of customers.

All of these links are part of one common process with one common goal: serving the customer. What a company does in one stage of the process can affect what happens in many other stages. Manufacturing affects marketing, and marketing affects sales. If any link in the chain is broken, the primary goal of the chain—serving the customer—goes unfulfilled.

A chain is only as strong as its weakest link, so companies must pay attention to every link. They also must maintain strong connections between the links. Different departments must talk to one another and work with one another. If a company fragments into a bunch of loosely connected fiefdoms, it will lose out to a more coordinated competitor. The problems at Xerox's Palo Alto Research Center, discussed earlier, provide an example. The researchers in Palo Alto were top-notch, but they rarely talked to other groups within Xerox. Thus, the product–customer chain was broken, and products got to market late, if ever.

Another important marketing chain involves what are known as "consumption patterns." These patterns are, in fact, product chains: they link together the sales of different products.

Computer companies generally don't develop their own applications software, nor do most provide customized product packages or localized service and support. These activities must be done through third parties. Yet the customer looks to the computer supplier to provide the whole product. The computer company, if it is customer driven, will organize and coordinate all the third-party suppliers in order to assemble the whole product package for the customer.

When infrastructure-pattern chains are broken, trouble is sure to follow. If one link of the chain, say, software, is missing, the products and services can't be used. Intel's success in the microprocessor business is largely due to its understanding of consumption patterns. Intel sells not only the microprocessors themselves but also the peripheral chips and development systems needed to put the microprocessors to use. Intel constantly adds new types of peripheral chips and microprocessors, and each new product enhances the sales of the others.

Distribution strategy also plays an important role in product chains. Sometimes all the products in the consumption-pattern chain are available, but are sold through different distribution channels. That can be just as bad for sales as having a missing link. The pieces all exist, but they are not linked together into a strong chain. Customers can't easily buy everything they need, so they might end up buying nothing. For this reason, retailers usually like to handle full product lines, not just individual hot products.

Another important chain is one linking different markets. Sales of a product in one market influence sales of the same product in other markets. In the personal computer business, for instance, the home and office markets are strongly linked. People who use personal computers at work are more likely to buy them for their homes. The reverse also holds: People with computers at home push for greater use of computers in the office. In many cases parents buy home computers for their children; then they, too, become interested in the machines. Before long, they want to use computers in their businesses. Rather than refer to this as the consumer market, I call it the home–business market.

The education market is another link in this chain. Children who use computers in schools often pressure their parents into buying home computers—usually the same brand used in the school. The university market is also important. Today's college students are tomorrow's decision-makers in the business world. In a few years they could be deciding what types of computers to buy for their businesses. They are likely to buy the same brand of computer they used in college.

Computer companies are now scrambling to take advantage of this linkage, offering computer systems at great discount to schools at all levels. IBM and Apple have donated thousands of computers to elementary schools, high schools, and colleges. Companies do this because they recognize the important effect the linkages between the school, home, and, eventually, the business marketplace can have on sales. Rather than think of the home as a consumer market, it is better to think of it as the home–education market. The home–office–school chain is important to the long-term success of most computer suppliers.

Competitor 8: The Product Concept

What do IBM, AT&T, CBS, Dow Jones, and Apple have in common? Five years ago the answer would have been not much. IBM sold big computers and office equipment. AT&T was in the telephone business. CBS was a television network. Dow Jones was a publishing company. And Apple sold personal computers. Today, however, all five companies compete against one another, at least indirectly. All are involved in the information business. All offer equipment and services that enable customers to access information quickly and efficiently. In the future they will compete directly with one another more often.

In this type of environment, companies can't afford to think about their products too narrowly. They must look for opportunities in—and expect competition from—every possible direction. A company with a narrow product concept will move through the market with blinders on, and it is sure to run into trouble. The product concept itself will become a competitor.

Earlier I mentioned the classic business school example of the decline of the railroads. Had the railroads considered themselves transportation companies, rather than railroad companies, they might have moved into the airline business. Instead, the railroads stuck to their narrow product concept and watched the new airline companies steal much of their business.

The same situation exists now in many evolving industries. Dow Jones, for instance, does not think of itself as a news service or a newspaper company. It sees itself as an information company. It must provide information in whatever form customers desire, whether written on paper, broadcast to radios, or sent over telephone lines to computer screens.

Similarly, a personal computer company should not view its product simply as a box with a keyboard and a display. If it sees its product that way, the company will have a narrow view of its competition. It will see other personal computer companies as its only real competitors, and it will plan its strategies with a false sense of the market.

In fact, many different products could compete with personal computers. Application-specific devices, such as pocket pagers and stock-quotation devices, are potential competitors. So are computer terminals and touch-tone telephones. Many companies are setting up information networks that allow users to access information through inexpensive "dumb" terminals. If people use these networks often, they might buy just a terminal, not a personal computer.

Home televisions also will be competitors. In two-way cable and satellite television systems, subscribers can use their televisions to get various types of information services. They can order airline and theater tickets, check their bank account balances, pay their utility bills, and check stock prices. Televisions will become even more formidable competitors to personal computers as manufacturers begin to build computers right into the sets. In 1990, Radius, a graphics-systems company in Silicon Valley, introduced a product called Radius TV. The product allows computer users to open a window on their computers and receive direct-broadcast, enhanced color television. Users can simultaneously watch CNN and place buy and sell stock orders with their brokers. Government officials can monitor the activities on the floor of the legislature by watching C-SPAN while they write memos or edit legislation. Products like this will change the definitions of the computer and television industries.

In developing their marketing and positioning strategies, personal computer companies must consider all these new competitors, and try to anticipate other challengers. If they limit their product concept and keep their blinders on, they are sure to be blindsided in the marketplace.

Competitor 9: Things that Go Bump in the Night

No matter how well a company understands its market, it is bound to be taken by surprise sometimes. New technologies, new companies, new applications all can shake up an industry when they emerge with little or no warning. I call these unanticipated events "things that go bump in the night." Companies don't see them coming. But like the iceberg that sank the *Titanic*, they can do a lot of damage.

There are more things going bump in the night today than ever before. The prime reason is the speed-up in technological innovation. According to one estimate, 99 percent of all technological innovations in the history of mankind have occurred in the past twenty years. Each year more innovations come about than appeared the year before, and each innovation holds the potential for shaking up a company, if not an industry.

The base of scientific knowledge, from which technology evolves, is continuing to grow rapidly. According to some estimates, more than 90 percent of all the scientists that ever lived are alive today. More important, scientific knowledge is being put to use more quickly than ever before. Engineers are constantly shrinking the time it takes to translate scientific advances into new technological products. One researcher, studying twenty major innovations, found that the time lag between scientific discovery and technological product has dropped by 60 percent since 1900.

No company in a technology-based industry is safe from unanticipated bumps in the night. The steel industry, the petroleum industry, even the textile industry all have been jolted by technological change. A decade ago the major pharmaceutical companies thought their industry was fairly mature and stable. Then came the development of recombinant DNA technology, and now dozens of new companies are challenging the products of established industry leaders.

The semiconductor industry has been predicting a major shakeout for fifteen years now. In the late 1960s many industry experts predicted that the semiconductor industry would soon resemble the auto industry, with only three or four leading manufacturers. They argued that the business was too capital intensive for new companies to join. At the time there were about ninety-five semiconductor companies. But today there are more than 200.

Steve Jobs, founder of Apple, made a similar prediction. He said the personal computer industry was too capital intensive to support new entrants, and he predicted a major shakeout in the industry. Even so, he left Apple and founded another company, NeXT. Certainly there will be shakeouts, but there is no way the industry will consolidate to a handful of companies, at least not in the near future. There is still plenty of room for technological innovation, and that means plenty of room for new competitors. "Shakeout" and "consolidation" are two words I try to avoid using. They make managements believe that their industries are becoming less competitive. With few players, the marketing task is easier. Few industries can claim less competition this year than existed five years ago.

There is no way for companies to avoid bumps in the night. But companies can be prepared for them. They can stay humble, expect the unexpected, and react quickly when the unexpected occurs. They can stay close to customers and build solid relationships with all members of the industry infrastructure, including their customers. Change is a market phenomenon. Managements must understand that no company is too big, and no industry is too capital intensive, to be shielded from the havoc caused by technological innovation.

Competitor 10: Yourself

This competitor is the toughest of all. Machines and products don't compete, people do. People have to spot the market opportunities and take advantage of them. People have to develop the products and competitive strategies, and allocate resources and develop customer relationships.

There are many ways people end up competing with themselves. When people underestimate their own ideas, just because the ideas have never been tried out before, they are competing with themselves. When, on the other hand, people develop an air of omnipotence and believe they can't fail, they also are competing with themselves. When people are unwilling to listen, when they are unwilling to change, when they are unwilling to experiment, they are competing with themselves.

People must leave themselves open to think creatively. With markets changing so rapidly, managers must be able to analyze new situations and apply creative approaches to them. Old approaches to new problems simply won't work.

Above all, managers must pay attention to their customers. They must listen and respond to them. They must not underestimate their competition—or overestimate it. And they must continue to experiment. Successful companies are lead by people who are never satisfied being second best. Leaders make things happen. They have an attitude, a way of thinking that permeates the company. If managers adopt this pattern of thinking, this frame of mind, they can avoid the biggest problem of all: turning themselves into a competitor.

The rate at which individuals and organizations learn may become the only sustainable competitive advantage, especially in knowledge-intensive industries.

Ray Stata
Chairman and President, Analog Devices

Chapter 11 | # The Long Road to Success: The Macintosh Story

THE LAUNCH

Annual meetings for Fortune 500 companies are usually rather boring affairs. The corporate secretary announces the predictable results of proxy votes. Other executives read off long lists of corporate accomplishments and financial results. If the corporation has performed well in the past year, the stockholders applaud politely. If not, they ask a few questions and worry about their dividend checks.

But when Apple Computer stockholders met on January 24, 1984, they were not in for an ordinary annual meeting. More than 2,500 people jammed into the Flint Center in Cupertino, California. The atmosphere resembled that of a carnival and a revival meeting. Like a minister at the pulpit, Apple chairman Steve Jobs preached to the gathered masses. He told them that January 24 marked the beginning of a new revolution in personal computing. The cheers from the crowd rose to a crescendo as Jobs walked over to a table carrying a suitcase-sized case. Jobs unzipped the case, and the world got its first look at the Macintosh computer. The crowd screamed its approval.

Apple backed up the Macintosh introduction with an enormous media campaign. It ran twenty-four Macintosh commercials during the Winter Olympics, and it paid for a twenty-page advertising insert to appear in eight magazines. To build up media interest, Apple sent Macintosh press kits, including Macintosh T-shirts, to 3,000 financial analysts and journalists. Articles about Macintosh seemed to be everywhere. The computer appeared on the cover of thirteen magazines. Television news shows ran features about Apple and its revolutionary new product. So did *Time* and *Newsweek*. Computer dealers were besieged with inquiries and orders. Within 100 days Apple sold more than 75,000 Macintoshes.

To the outside world the marketing of Macintosh seemed like a big promotional event. But success does not come that easily. Hidden behind the splashy introduction were hard years of marketing development and planning. And in reality Mac marketing, or the adaptation of the product to specific markets, had hardly begun.

The story of the Macintosh success presents an instructive example of how to market new products in an era of rapid change. It is a story of a small group of people who succeeded in positioning a new product while ignoring many traditional rules of marketing. The members of the Macintosh team relied to an extraordinary degree on their collective intuition about the market, performing hardly any statistical analyses. They succeeded by constantly refining and readjusting the positioning of their product to match the ever-changing market environment.

Apple struggled to establish a coherent strategy for use when launching a new standard in the personal computer business. Apple's struggle demonstrates how products, companies, and markets continually change and evolve on the way to success. However, I caution the reader that the Mac was launched at a particular time and in a particular environment. What worked then probably wouldn't work now. The lessons to be learned by the Mac history are in the Mac's constant change and evolution.

THE BEGINNING

The story of Macintosh begins nearly five years before the product's introduction. In 1979 Jef Raskin, an engineer who had written technical publications and manuals for Apple, came up with the initial idea for Macintosh. While many engineers were trying to modify or enhance the enormously successful Apple II computer, Raskin believed that Apple needed a radical new approach to computing.

The Apple II was itself a radical product in the world of computing. Introduced in 1976, the Apple II created an entirely new product category, that of personal computers. By 1980 Apple held an 80 percent share of the personal computer market. It had changed forever the way people thought about computers. Computers were now light enough to carry and inexpensive enough to sell through retail stores. Across the country, thousands of software developers dedicated themselves to designing programs for the Apple II, opening up myriad new applications for personal computers in homes, offices, and factories.

But Raskin's intuition told him the technologies used in the Apple II would never bring computing to the masses. The Apple II was fine for hobbyists and people willing to spend long hours learning arcane computer commands. But Raskin wanted to build a computer that anyone could use. At first he called it "Everyman's Computer." Raskin knew that the Apple II and its direct descendants would never fill that role. Raskin didn't do any market research or conduct customer surveys; he relied on intuition. He knew that Everyman's Computer would have to be far easier to use than the Apple II or any other computer then on the drawing boards.

To make his computer easier to use, Raskin relied on a new user interface—that is, a new way for users to interact with the machine. Communication with the machine was based on pictures more than on words. Using a hand-held device called a mouse, users would point to pictures that appeared on the screen in order to tell the computer what to do next. The result was a much more natural interaction between user and machine. Rather than type out strange commands, as they had to do with other computers, Macintosh users would simply point. The screen, for instance, could simulate a desktop. Users could point at items on the screen just as they would pick up pieces of paper from their desks.

The original designs for Macintosh were quite different from the final product. Raskin and his colleagues figured that the machine would weigh less than ten pounds. In fact, it ended up weighing seventeen pounds. They hoped that the machine could be priced under $1,000. Instead, it came to market at $2,495. The original design was battery operated. The final product was not. One thing about the Macintosh project never changed, though: the commitment to making Macintosh easy to learn and use.

Raskin's dream computer probably never would have been more than a dream had it not been for Steve Jobs. When Jobs saw Raskin's design, he fell in love with it. Like Raskin, Jobs believed that Apple had to try some radical new approaches to computing. Jobs had tried to get involved in Apple's Lisa project, but was not really welcomed there. He was a founder and a top officer of Apple, but in many ways Jobs was a man without a company. Macintosh captured his imagination. He became head of the Macintosh project.

Under Jobs, the Mac project became a company within a company. In effect, Jobs was heading a start-up company, just as he had done in the early days of the Apple II. The Mac group remained rather small. In early 1982 there were still only twenty-five members.

In the rest of Apple the Mac project had little credibility. Few people expected it to succeed. But inside the Mac project everyone was a believer. Mac team members shared a vision of what personal computers could and should be. They developed an extraordinary sense of community and camaraderie. They often would sit on the floor and talk through the night. Their intuitions were strong: They knew they had a winner.

THE AUDIENCE

During the first two years of the Macintosh project, there was no marketing staff involved. The project was driven purely by technology. The Macintosh team and its budget were both small, and team members wanted to put all their time and money into development, not marketing.

In 1982, though, the Macintosh group began adding some marketing people and holding strategy sessions. Four or five people (including myself) met every two weeks to talk about positioning for the Mac. The meetings were mostly brainstorming sessions. We spent a great deal of time trying to figure out how to create the market, what users to target, who was going to use Macintosh, and how we could best communicate new ideas to those potential users.

We decided that Mac's target audience would not be a traditional market segment. For some products, it is possible to target a neat slice of the market—for example, all small businesses in service industries with annual sales between $500,000 and $1 million. But that type of segmentation wouldn't work for Mac. Mac cut across the usual boundaries. We needed to come up with what I call a "concept market." Most marketing managers look to divide a market along demographic or geographic lines. But concept markets are divided along "psychographic" lines; that is, they include people with similar attitudes and beliefs.

We came up with the idea of targeting "knowledge workers." These are people who typically sit at a desk during the day. They create ideas, make plans, analyze data. Knowledge workers exist in many different settings. Some work in large offices as professionals, others in homes as consultants, and still others in dormitories as college students. One internal Apple marketing plan described knowledge workers this way: "Knowledge workers are professionally trained individuals who are paid to process information and ideas into plans, reports, analyses, memos and budgets. They generally sit at desks. They generally do the same generic problem solving work irrespective of age, industry, company size, or geographic location. Some have limited computer experience—perhaps an *introductory* programming class in college—but most are computer naive. Their use of a personal computer will not be of the intense eight-hour-per-day-on-the-keyboard variety. Rather they bounce from one activity to another; from meeting to phone call; from memo to budgets; from mail to meeting. Like the telephone, their personal computer must be extremely powerful yet extremely easy to use."

After making some rough calculations, Apple figured there were about 25 million knowledge workers in the United States that might use Macintosh computers. That included 5 million in small businesses (those making less than $5 million in annual sales), 5 million in large businesses (Fortune 2000 companies), and nearly 9 million in medium-sized businesses. The other knowledge workers were in the college and home markets.

At the time, not many knowledge workers were using personal computers. People need a significant amount of expertise and training to use the Apple II. Selling the Apple II was like selling a telephone that the buyer had to put together. But Mac would be different. It would be nonintimidating and easy to use. People could learn to use it in two to four hours, rather than in the twenty to forty it took to learn the Apple II. Knowledge workers would feel comfortable with it. With Mac, Apple could tap into a new market of 25 million workers. Mac could become the standard product of knowledge workers.

We began to view knowledge workers as those who would be the next stage in the adaptation sequence for personal computers. As discussed in Chapter 6, most new products are accepted by the market in stages: first by Innovators, then by Early Adapters and Late Adapters, finally by Laggards. Before Macintosh, personal computers had been purchased only by Innovators. Innovators were willing to read a 400-page user's manual and spend twenty to forty hours learning how to use the computer. The computer was an important part of their lives.

But the market had begun to run out of Innovators. In the same way many Americans had waited for Interstate 80 before heading West, so too were many people waiting for easier-to-use computers. Most knowledge workers were part of this group. The next generation of computers had to address the needs of these workers.

We spent hours discussing such questions as: Who are knowledge workers? Where are they? How can we identify them? The strategy was to make Mac so unique and innovative that knowledge workers would be romanced into using it.

At the marketing meetings we also spent a lot of time thinking about words and language. How could we communicate to the public about Mac? We decided we had to create our own vocabulary. Revolutions create their own language. We knew language would be very important in positioning

Mac. We knew Mac was radically different from other personal computers on the market, and we didn't want them compared. We didn't want people to compare operating system versus operating system, keyboard versus keyboard. Instead, we would create our own words. Then, when the press and dealers talked about computers, Mac would stand out. It was a type of forced differentiation.

Some words were obvious, like _mouse_ and _user interface._ But most of the discussion focused on the word _appliance._ Mike Murray, the marketing manager for Mac, felt that Apple could position Mac as an information appliance. Some of us, myself included, were highly skeptical. We argued that computers were more complex than refrigerators and other traditional appliances. We also worried that the word _appliance_ would target Mac toward the home, not the office.

But Murray insisted. He gave away miniature food processors to members of the Mac team. In August 1982, when Murray wrote the first product plan for Mac, he drew heavily on the appliance metaphor. But the marketplace didn't pick up on the metaphor and the analogy died a quiet death.

Differentiating Mac from competitors' products was only part of the marketing problem. Just as important, the Mac marketeers had to figure out a way to differentiate Mac from the rest of the Apple product line. All Apple computers, those on the market and those in development, were aimed at roughly the same audience. There were serious fears that each Apple computer would cannibalize the others.

All companies with more than one product face this internal conflict. The typical solution to this problem is to divide up the turf. Target one product at small businesses, another at professionals. Sell one product at a premium price, another at a discount. If products overlap too much, dealers and customers will be confused. Dealers won't know how to sell the products, and customers won't know which one to buy. Each product must give up a bit of its territory for the sake of maintaining its clarity.

At Apple, however, this normal approach seemed impossible. Each Apple product group was an individual fiefdom, and the managers of each group were more interested in competition than cooperation. Many Apple products were targeted at small businesses and professionals, but the different Apple marketing groups wouldn't sit down together to resolve the conflicts.

Perhaps the biggest conflict occurred between the development of Mac and the development of Lisa. The two projects had very different roots. The Lisa project was a major corporate effort, with lots of people and lots of money available, while the Mac project involved much less cooperation and fewer resources. By 1982 the two projects were on a collision course.

Lisa's development group hoped to establish its product as one that would set a new standard in personal computing. Mac's group had the same hopes. Lisa's team planned to make its computer dramatically easier to use than traditional computers through the use of a new graphics-based user interface. Mac's team had the same plans. Lisa used the Motorola 68000 microprocessor. Mac used the same. Lisa used a mouse as a pointing device. So did Mac.

Even the target audiences were alike. In the beginning, Mac's designers shunned business applications. They wanted to bring computer power to the masses, not to money-hungry businesspeople. They were suspicious of anyone who read the *Wall Street Journal* and actually enjoyed it. But when Jobs took over the Mac effort, he steered the product toward office applications and knowledge workers—aiming it at much the same audience at which Lisa was aimed.

The two groups became intensely competitive. Each wanted to outshine the other. The general managers of the two groups, Jobs and John Couch, even made a personal bet over which computer would reach the market first. The loser would have to throw a celebration party for the winner.

Conflicts also existed between Mac and the aging Apple II. Most people saw the Mac as the replacement for the Apple II. The Mac would make the Apple II obsolete. That could cause big problems for Apple, however. Despite its incredible growth, Apple was still, to a large extent, a one-product company. It was doing $500 million in sales with one basic product, the Apple II. What would happen if that one product became obsolete overnight?

When developing a successor to an existing product, companies usually try to arrange for one product to replace the other gradually. As sales of the new product shoot upward, sales of the old one gradually fall. Many Apple marketeers worried that the transition from the Apple II to the Mac would not be so smooth. When Mac was introduced, they figured, Apple II sales would plummet. But Apple would have to build up its production of Macs slowly. The transition could be disastrous.

MACTROUBLE

The year 1983 was one of turmoil and change for Apple Computer. It started smoothly enough. In January Apple introduced its Lisa computer. (Lisa had moved through the development cycle faster than Mac, allowing

John Couch to collect on his bet with Jobs.) The marketplace quickly acknowledged Lisa as a revolutionary product. Magazines ran glowing reviews about the new machine. Some people complained about Lisa's $10,000 price tag, but nearly everyone agreed that Lisa set a new direction for the personal computer industry.

The enthusiastic reception to Lisa was due, in part, to a remarkably favorable market environment. Everything seemed to be going Apple's way. The personal computer industry as a whole was still expanding rapidly. Almost everyone was prospering—Atari, Osborne, Commodore. *Time* magazine had even selected the personal computer as its "Machine of the Year" for 1982.

The entry of IBM into the personal computer market in 1981 had scared some competitors, but IBM's presence seemed to be helping rather than hurting other manufacturers. IBM's stamp of approval gave added credibility to the new industry. The rising tide of enthusiasm about personal computers lifted all boats in the industry.

No one benefited from the booming market as much as Apple. Apple was still the clear industry leader in sales and profits. Just five years after Steve Jobs and Steve Wozniak started the company in a garage, Apple was a member of the elite Fortune 500. Never before had a company joined the ranks of the Fortune 500 so quickly.

But in the first six months of 1983 the market environment for personal computers shifted dramatically. Prices continued to drop and competition continued to grow, putting the squeeze on industry profits. The personal computer industry was no longer a paradise. One company after another ran into financial trouble. Osborne and Victor filed for bankruptcy. Texas Instruments and Atari lost hundreds of millions of dollars.

Apple, too, began to struggle. Its quarterly profits fell for the first time in its history. Even worse, the heralded Lisa computer faltered in the marketplace. Production problems forced Apple to delay shipping the new computer, and Lisa never regained its initial momentum. Software was slow in coming to market. Also, the price tag dictated that the Lisa had to be sold to sizable corporations. But Apple had no experience selling to corporate America.

Apple had expected to sell lots of Lisas to knowledge workers at large corporations, but the shifting environment foiled Apple's plans. Before 1983 personal computers were seen as stand-alone machines. Individuals in large corporations bought personal computers without even consulting the manager in charge of management information systems. Apple hoped to sell

Lisas to those people. It figured people would be willing to spend $10,000 for a technologically advanced stand-alone system.

Things didn't work that way. MIS managers began to worry about the uncontrolled influx of personal computers. Many corporations were getting stuck with a random collection of machines from many different manufacturers. In most cases these machines were incompatible. They didn't use the same software or peripherals, and they couldn't be connected into networks to share data and information. MIS managers declared war against these unauthorized computers. They sought to regain control over all computer resources in the company.

That was bad news for Apple and good news for IBM. MIS managers had bought from IBM for years, and they trusted IBM. They began to buy IBM personal computers by the hundreds, even thousands. Apple couldn't get its foot in the corporate door. MIS managers saw Lisa as too expensive and lacking in networking and data-communications capabilities. And the Apple II simply wasn't powerful enough for many new applications.

IBM continued to gain momentum in the industry. Its personal computer market share grew from 18.4% in 1982 to 30% in 1983. IBM seemed invincible. Rumors circulated about IBM's next personal computer, an inexpensive home computer named Peanut. Analysts and retailers predicted that Peanut would be the first true mass-market computer. They declared Peanut a winner even before it was introduced. Stock prices of other personal computer companies fell sharply in anticipation of Peanut. Apple's Macintosh seemed out of step with the environment. Apple was getting ready to introduce a $2,500 computer while everybody in the industry was talking about the $500 Peanut.

In the span of a few months IBM's image in the industry had changed dramatically. To customers and analysts, IBM was the dominant industry leader. To other manufacturers, IBM was now Enemy No. 1. When IBM introduced its PC, Apple had run an ad in the *Wall Street Journal* saying "Welcome IBM." But now Apple wanted to take in the welcome mat. A joke circulated around Apple's Cupertino headquarters: Question: What are the two biggest lies in the world? Answer: The check is in the mail, and Welcome IBM.

COMEBACK NO. 1

When John Sculley joined Apple as chief executive, in April 1983, he faced a desk full of troubles. With the disappointing sales of Lisa, Macintosh was now more important than ever. It was no overstatement to say the future of Apple depended on Macintosh.

Apple managers were still enthusiastic about Macintosh, but they were depressed about the market. The environment was nasty. No one knew how the market would accept Macintosh, and no one was quite sure how to market Mac. The marketing plan for Mac had been kicked around for a few years as the marketplace constantly changed. Somehow, Apple had to fit Macintosh into the current environment.

Sculley also had to clean up two other problems. First, he had to make sure the company remained profitable. Apple had built up its facilities for huge Lisa sales that never came. The company was overextended financially and profits were falling. It would be disastrous for Apple if the financial situation ever dipped into the red. The company would lose its credibility and its corporate positioning. In people's minds, Apple would be in the same category as Osborne, Atari, and Texas Instruments. People would start questioning Apple's future. But if Apple could cut back on expenses and remain profitable, even with lower sales, it could stay separate in people's minds from the Osborne-Atari crowd.

The second task was to rescue Lisa. Product success builds on product success. If a company has a failure, it loses credibility for its next product introduction. If Lisa remained problematic, retailers and other people would keep worrying about it, thus diverting attention from Macintosh. To revive Lisa, Sculley and Jobs put together a plan to cut Lisa prices, increase the number of retailers selling Lisa, and develop some form of compatibility between Lisa and Mac so that the two computers could form a coherent product family.

While this was going on, Apple began to get some good news in the marketplace. The latest version of the Apple II, called the Apple IIe, was selling very well. During 1983 Apple sold more than 700,000 Apple II computers, up from approximately 300,000 in 1982. Apple marketing managers began to realize that the Macintosh might not make the Apple II obsolete after all. Year after year, sales of the Apple II continued to rise. IBM was already in the market and Mac was in the wings, but people kept buying the Apple II. The large installed base and the third-party software suppliers kept the Apple II healthy. Apple marketeers began to believe there would be life after life for the Apple II.

Macintosh and the Apple II, once seen as competitors within Apple, began to differentiate themselves. A few years earlier, Apple managers figured that the Macintosh and the Apple II both would sell at price points of between $1,000 and $1,500. But their price points were now drifting further apart. While the price of the Apple II continued falling, the development costs of Macintosh continued to rise. Macintosh would have to sell at $2,000 or $2,500, well above the Apple II price.

Also, Apple began to split the target markets for the Apple II and the Mac. The Apple II was targeted more toward the education market and other vertical markets. Since the introduction of the Apple II in 1976, software developers had written thousands of specialized programs for the computer, so it could be used in thousands of specialty applications. There were Apple II programs for use in managing pig farms and others to help educate kindergarten children. Even after the introduction of Mac, these applications would not disappear. Macintosh, meanwhile, would be aimed at the horizontal market of knowledge workers.

At about the same time, the Macintosh project received some help from a most unexpected source: IBM. After two years of flawless performance in the personal computer market, IBM began to make a few mistakes.

For one thing, the IBM Peanut, officially called the PCjr, fell short of expectations. With its toy-like keyboard and limited memory, the PCjr disappointed retailers and customers. People began to realize that IBM was not invincible. It was not an automatic winner. Among retailers and software designers there was a new hesitancy, an uncertainty, about IBM. The computer giant would have to reestablish its credibility in the personal computer market.

What's more, the PCjr flop gave Apple more flexibility in the positioning of Macintosh. Had PCjr been a raging success, people inevitably would have compared Mac to PCjr, even though they were totally different machines aimed at very different markets. Positioning Macintosh properly would have been difficult. When PCjr fizzled, that problem disappeared.

Even before the PCjr flop, some retailers and software developers had begun to worry about their relationship with IBM. Many retailers had become heavily dependent on IBM. At some stores IBM products accounted for 75 percent of all sales. No retailer likes to be that dependent on a single supplier.

IBM soon began expanding its own distribution channels. It added more IBM Product Centers and expanded its direct-sales force. Analysts reported that IBM was expected to sell 60 percent of its computers through its own sales channels, up from 40 percent in 1982. Independent retailers were justifiably nervous. What was to stop IBM from selling 70 percent or 80 percent or even 100 percent of its products through its own channels, leaving little or nothing for independent retailers?

Our external audits revealed that the independent software developers also were becoming heavily dependent on IBM. They made most of their money selling programs designed to run on IBM personal computers. These software designers had helped make the IBM PC a success, but now IBM was beginning to compete with them. IBM was publishing more and more software itself, taking sales away from software companies. Some software companies licensed their software to IBM, but that arrangement was a mixed blessing. They got the licensing fees, but lost the market.

In addition, it was widely rumored and discussed in various analysts' reports that IBM was in the process of developing its own operating system. The introduction of this system would mean that the control of the industry would return to the proprietary IBM architecture. Independent developers of application programs would then be at the mercy of IBM. Major software companies were clearly feeling uneasy about the trends in the industry.

All of this activity created a new environment in the personal computer industry. The concept of Fear, Uncertainty, and Doubt had been turned on its head. In most markets, IBM uses FUD to its advantage. Customers are fearful of buying from any supplier other than IBM. Buying from IBM is the safe bet in an uncertain world.

But in the personal computer market, IBM had FUDed itself. The company itself was the source of uncertainty and doubt. Members of the industry infrastructure—resellers, third-party software designers—no longer trusted IBM. They were suspicious of IBM's motives, and uncertain of IBM's future directions. They still wanted to do business with IBM, but they didn't want to be too dependent on the computer giant. Our external audits, conducted within the infrastructure, uncovered the depth of the fear that existed. The president of a large retail computer chain told us that although IBM products made up 95% of his business, he expected that within five years IBM would be his biggest competitor. Fred Gibbons, president of Software Publishing, the developer of the PFS family of software products, told me he was "sweating bullets" over how to deal with the changing business conditions.

THE INFRASTRUCTURE

This new market environment presented a tremendous opportunity for Apple and its Macintosh. Members of the industry infrastructure no longer held IBM in the highest regard. They were looking for alternatives to the IBM PC. Apple recognized this, and set out to turn members of the infrastructure into Mac supporters.

The industry infrastructure is enormously important in the personal computer business. No personal computer, no matter how powerful it is, no matter how advanced it is, can be successful in the marketplace without the support of the infrastructure. Software designers must write programs for the computer, resellers must develop myriad vertical markets, and analysts must praise the machine in their newsletters.

If a product can win the support of the infrastructure, it has a better than average chance of winning in the marketplace. The infrastructure system works like a chain reaction. Confidence about the product's success spreads by word of mouth, and enthusiasm grows. If the leaders in the software industry support the new machine, resellers are more likely to recommend it to customers. If the initial customers reference the product favorably, analysts and journalists tout it as a winner. The product builds momentum and gains credibility. To customers confused by new technologies and changing markets, such a product looks like a safe bet.

Months before the Macintosh introduction Apple began cultivating favorable relations with the infrastructure. As word spread about IBM's plans to sell 60 percent of its computers through its in-house channels, Apple solidified its own relations with dealers. It set up regional dealer councils to serve as liaisons between itself and the retailers. And it cut back its plans for a direct-sales force, making a commitment to independent dealers that they would remain the primary sales channel for Apple computers.

Apple also went after software developers. With these people it had some fences to mend. At one time Apple had been the darling of the software industry. Everyone had wanted to design software for the Apple II. But Apple had turned arrogant when it became successful. When it developed the Apple III and Lisa, Apple didn't let software companies work with the computers before the formal product introductions. Apple believed that it could develop the systems and applications software itself rather than let others make money off its products. Software designers grew frustrated with Apple, especially when the Apple III and Lisa fell short of market expectations. Software companies began focusing on the IBM PC market instead.

Now Apple tried to turn that around. With new humility, Apple managers and engineers visited software developers and asked them to develop programs for the Macintosh. Apple offered to help and support them in their development efforts. About 100 companies signed up, including three of the biggest and most influential—Microsoft, Lotus, and Software Publishing.

Each software company signed a nondisclosure agreement. But that didn't stop word of the new product from spreading. Although the Mac was not the easiest product to develop software on, its bit-mapped graphics and user interface allowed software designers to do things that couldn't be done on the Apple II or IBM machines. Developing in the Mac environment appealed to the highly creative software community. That community is small, and everybody talks to everybody else. Before long, everybody was talking about the Mac. Designers heard that Microsoft and Lotus were working on Mac software, so they assumed Mac must be a winner. They, too, wanted to get in on the action. Everybody wanted to design software for the Mac.

Dealers like to hear about that type of commitment directly from top management, so John Sculley traveled around the country and met with Apple dealers. By the time of the introduction, about 4,000 dealers had been trained on how to use the Macintosh. Many had fallen in love with it. Apple won shelf space for the Mac. More important, it won space in the dealers' minds.

Finally, Apple took its message to analysts and industry luminaries, and later to journalists. Key members of these communities got seven-hour demonstrations of the Macintosh, with plenty of hands-on time. Many of these people are computer aficionados, and they fell in love with the Mac as soon as they began playing with it. Before long, they started to spread the good word about the new product.

The initial task was complete: The infrastructure was lined up solidly behind Macintosh.

MACMESSAGES

In preparing its merchandising and public relations campaigns, Apple marketeers had one overriding goal: They wanted to establish Macintosh as the third standard in the personal computer industry.

Apple argued that only two products had emerged as industry standards in the eight-year history of the personal computer industry. Those products were the Apple II and the IBM PC. The Mac would not become a new standard overnight, but Apple wanted to plant the idea early.

To convince customers and the media that Mac was indeed a new standard, Apple stressed its product features. Mac marketeers wanted to drive home the point that Mac was significantly different from other personal computers. They identified four key messages about the Mac. Then they repeated those messages over and over. The messages were:

Mac offers "Lisa technology."   Although Lisa had fallen short of expectations in the marketplace, its technology won rave reviews. People were intrigued and impressed with Lisa's friendly user interface—its mouse pointer, its pull-down menus, its bit-mapped graphics, its windowing capabilities. Apple wanted people to know they could get these same features in Macintosh. Or, as Apple marketeers love to say, Mac offers "radical ease of use."

Mac uses a 32-bit processor. Many people don't understand what "32-bit" means, but they know it stands for advanced technology. After all, the processor in the IBM PC is only a 16-bit processor. As one piece of Mac literature put it: "Mac offers incredible power under the hood."

Mac offers personal-productivity tools. This is where Apple's infrastructure development paid off. At the time of the Mac introduction Apple could boast that 100 leading software firms were working on Mac software. That software would increase personal productivity and creativity.

Mac comes in one box. This message helped eliminate customer fear. Mac is compact and simple. It can be taken out of the box and plugged in. Also, the computer is easy to carry around. As the marketing plan stated: "Macintosh fits comfortably on your desk and in your life."

The Mac marketeers spread these four messages everywhere: in meetings with the media, in meetings with dealers, in customer brochures. At the time of the product's introduction, every one of the 10,000 salespeople selling the Macintosh could recite the four key messages. Apple kept its messages clean and simple. With the Apple IIe, Apple gave forty pieces of information to dealers. With Mac, the company gave dealers a single book.

In meetings with journalists, Apple added a number of other messages about Macintosh. To some publications, Apple stressed the new automated Macintosh factory. The factory, filled with the latest robotics equipment, would turn out one Macintosh every twenty-seven seconds. Many people were worried about the manufacturing capabilities of American companies. The United States seemed to be losing out to Japan in robotics and other manufacturing technologies. The Mac factory story stood out as a bright spot in this dark environment. While Atari had just shifted its manufacturing overseas, Apple was bringing its manufacturing back to the United States. What's more, the Mac factory was located in Fremont, California, where General Motors had just closed down an automobile factory. The opening of the Mac factory was seen as a clear case of a new industry taking over from an old one. It helped give Macintosh a higher profile. Mac was not just another computer. It was a symbol of the American future.

The Mac marketeers also focused attention on the engineering team that developed the Mac. The team consisted of a dozen or so young people who had contributed their sweat and talent for Macintosh. They had worked day and night for four years. The story of the engineering team showed Apple as a human company, a personal company.

Apple already was perceived as having a strong corporate personality based on the success of the Apple II. The Macintosh story built on that image. Apple was a company the public liked to root for. Apple people were young, dynamic, and innovative. Steve Jobs and Steve Wozniak became models of entrepreneurial success in America.

With these and other stories, Apple turned the Macintosh launch into a huge media event. Mac managers began giving key journalists a sneak preview of the machine months before its January 24 introduction. In mid-January the journalists were taken on a big press tour, capped by a Macintosh "coming-out party" on January 22. On the day of the introduction Apple mailed out 3,000 press kits, each containing not only pictures and press releases but also a Macintosh T-shirt. After five years of being in development, Macintosh finally was moving outside of Apple and into the world.

MACADVERTISING

Apple's advertising strategy for Macintosh broke into two categories: the "1984" television ad, and everything else.

The "1984" ad, created by Chiat/Day Advertising, used images from George Orwell's classic book *1984*. It was unlike any ad that Apple—or any other company, for that matter—had ever done. It was shown nationally only once, during the 1984 Super Bowl. But it became the most-talked-about advertisement in years. Inside Apple it caused arguments, controversy, and, in the end, big smiles.

The ad looks like a clip from a movie or a rock video, not like a television commercial. It begins with a view of a dark and somewhat eerie room. Men with shaved heads sit on row after row of benches. They stare blankly at a huge screen on the wall, where a cold and grim man, clearly representing Big Brother, talks in a monotone.

Suddenly, the image shifts to that of an athletic young woman dressed in bright red running shorts and an Apple shirt. She is running down a dark corridor, carrying a sledgehammer. Chasing her is a group of uniformed men, apparently the Thought Police. As she enters the main room, where the men are, she swings the sledgehammer and throws it at the screen. The screen shatters and a huge gust of wind sweeps past the zombie-like men.

The screen goes blank, then the Apple logo appears. A narrator says: "On January 24, Apple will introduce Macintosh. And you'll see why 1984 won't be like '1984.' "

The ad, produced in London for $500,000, almost died before it reached the air. Mike Murray, Mac's marketing manager, showed the commercial to Apple's board of directors in November. Murray loved the ad and expected the board to love it too. He was in for a surprise. Murray explains it this way, "When the commercial was over, I thought I had just come out of a funeral. Their faces were extremely solemn. I thought that I had just made a major career decision. One of the directors looked at Steve (Jobs) and said: 'You really like that?' He was just incredulous. Another director had his head on the table and was pounding with his fist. At first, I thought he was laughing. But he was nearly crying."

Murray tried to explain the reasoning behind the ad. He believed that Apple needed an attention grabber. People in the personal computer industry already knew about Macintosh. But out in the marketplace Macintosh was a total unknown. If nothing else, the ad would establish the Macintosh name. Even if the commercial seemed a bit bizarre, or maybe *because* it seemed a bit bizarre, many of the 80 million people watching the Super Bowl would remember the Macintosh name. The commercial would make them sit up and take notice: Macintosh was something new and radically different.

In addition, the commercial would add to Apple's personality. It showed Apple as daring and creative, and it set the stage for the Apple–IBM battle. Many viewers would recognize Big Brother as a thinly-veiled image of IBM. The interpretation was clear. Apple, the daring and creative upstart, was taking on the colossus. Framing the battle that way gave Apple a big advantage. Americans like to root for underdogs. Everyone likes to see entrepreneurs succeed. It made sense for Apple to play up the Apple–IBM battle.

The board members didn't buy these arguments. They told the Mac managers to resell the airtime they had bought. Apple tried to sell the air time, which cost $950,000 per minute. But as luck would have it Apple couldn't find any buyers, and so the "1984" commercial went on the air.

Reaction to the commercial was extraordinary. Everyone in the advertising community talked about it. Newspapers ran articles about it. Television news programs showed the commercial as a news item, giving the Macintosh even more exposure. Inside Apple the doubters and skeptics, myself included, realized that the commercial had worked. At the next board meeting Murray got a standing ovation.

In the long run, the ad did more for revealing Apple's character than it did to create sales for the Mac. The ad widened the gap between MIS people and Apple. It told the world that Apple was revolutionary, but traditional computer buyers were anything but revolutionary. But if Apple was really serious about challenging IBM, where were the systems technology, software, support, service, and all the things it needed to create corporate information systems? The ad won awards but the Mac continued to struggle to gain entry into corporate America.

The rest of the Mac advertising took a very different approach. If the "1984" commercial appealed to the consumer on an emotional level, then the rest of the advertising appealed on a rational level. The ads, both print and television, were very product oriented. They presented product features and product benefits. And they made direct comparisons between the Mac and the IBM PC.

This approach was a switch from that used in Apple's advertising for Lisa. Apple had used lifestyle ads to promote Lisa. One television ad showed a young executive dressed in running clothes, working with his Lisa in his office. He was on the telephone with his wife, telling her he would soon be home for breakfast.

Apple learned a lesson from those ads. It learned that lifestyle ads simply don't work for marketing complex new products. When marketing new technologies, differentiation must begin with the product. Companies must start by giving tangible evidence of the product benefits. Companies can't just go out and say "We are the leaders." Intangibles, such as leadership image, must grow out of the tangibles. Corporate and lifestyle advertising start with the wrong things. They start with the intangibles.

With its Macintosh advertisements Apple shifted back to focusing on tangibles. A big chunk of the advertising budget went toward producing a twenty-page magazine insert, which ran in eight different magazines including *Fortune, Time,* and *Business Week.* The insert was filled with information about Macintosh. On one page, a cutaway diagram showed the inside of a Macintosh. On a four-page foldout, the ad showed how to use the Macintosh mouse. Point. Click. Cut. Paste.

The insert was enormously successful. It garnered one of the highest recall rates of any advertisement ever run. Apple printed up extra copies of the advertisement and used it as a point-of-sale brochure. The ad did its job well. It helped differentiate Macintosh from other computers and it solidified Macintosh's product positioning.

COMEBACK NO. 2

Computers are not born perfect. They never work fast enough, or have enough memory or software. They must adapt and be adapted to specific markets. That is true of most technical products. Once a product hits the market, the market begins to mold the product to suit its own unique ways of doing business. Mac's technical inflexibility cut off its access to and acceptance by many potential markets. The Mac user interface was a winner with every user. But the Mac did not have enough memory and there was no way to expand or adapt it for new applications. Apple forgot the lesson it learned with the Apple II. The Apple II had seven expansion slots inside, allowing hundreds of third parties to adapt it to different uses. The market infrastructure contributed as much to the success of the Apple II as Apple did.

The Mac's inflexibility also led to management disagreements and eventually to Steve Jobs' leaving Apple. Further, its futile challenge to take on IBM in the big business environment was arrogant. The "1984" ad made a statement about Apple but it didn't change Apple's market-share decline. Too often Apple managers relied on promotions and advertising to take the place of marketing. John Doerr, general partner of Kleiner Perkins Caufield & Byers, once remarked that "Apple is a vertically integrated advertising agency."

Apple began as a countercultural company that believed that the MIS managers of corporate America would be overthrown by a computer-user coup. Mac did enter many corporations at the user level, but as personal computers began to proliferate throughout the corporation, compatibility became a major issue. At the time, many people said that corporate America would not buy from a company whose people wore blue jeans and held meetings while sitting on beanbags. I never believed that blue jeans and beanbags were the issue. The problem was with the product and Apple's understanding of the changing marketplace.

The enthusiasm and excitement that often accompany the introduction of a product may help to get the product established. But that's not enough. Sustaining a product in the market requires constantly adapting the technology to the needs of the market. It means constant development of market segments. "Knowledge worker" was too broad a description for the members of Mac's targeted market. Teachers, engineers, scientists, salespeople,

PERSONAL COMPUTER ENVIRONMENT 1977

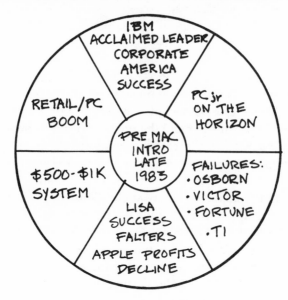

Figure 13.

PERSONAL COMPUTER ENVIRONMENT 1991

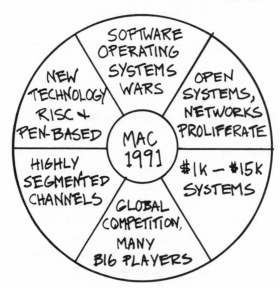

Figure 14.

production managers, and accountants are all knowledge workers, but each has different needs. Each requires different types of software, support, and service. Marketing channels are different and competition within each segment is different. During the Mac introduction Apple relied too much on its internally developed views of the world and cut itself off from a large part of the infrastructure.

John Sculley unified Apple. He reorganized the company by steering it away from having independent Mac, Lisa, and Apple II divisions and bringing the separate manufacturing, marketing, finance, and engineering departments together to function as one company. Rather than compete with itself, as it had been doing, Apple now had a product line of which each product had different market opportunities. Lisa was abandoned. What remained was integrated into the Mac operation. All this restructuring took time and a toll on Apple and its people. The result, however, is a better Apple.

In 1984 Apple began working with Canon and Adobe to develop a laser printer. The resulting Apple laser printer was as technologically sophisticated as the Mac itself. Canon provided the laser engine. Adobe, founded in 1982, developed a software product called PostScript that resides inside the laser printer. PostScript translates the computer-generated information into a language that describes the printed page to the output of the printer. The result is low-cost, easy-to-produce, offset-quality documents created right at the desktop. The laser printer wasn't a new idea, but combining the Mac's graphics capability with Adobe software was a real breakthrough.

The software community had another capability to build on. Aldus, founded in 1984, introduced a product called PageMaker. This software allowed Mac users to design and produce books, newsletters, custom-made letterheads, and professional-looking documents on their desktop machines.

The desktop-publishing market didn't exist before the technology and software were in place. Research firms didn't mention it in their forecasts. At first, some Apple managers were unwilling to commit resources to this market because, in the words of one Apple manager, "it's not a mass market." But the more the market was explored, the better the opportunity for sales in it looked. Laser printers began selling Macs, and software became available under the heading, "desktop publishing." Resellers began giving seminars on and packaging hardware and software into desktop publishing systems.

The emergence of desktop publishing gave Apple an opportunity to define and own a market. IBM had a tough time competing in this market because its machine simply could not compete with the Mac in graphics applications. Desktop publishing was Apple's Trojan Horse into corporate

America. There are many stories about executives receiving letters that looked like they were printed on an offset machine and asking how the letter was done. Documents began to look better, and they were easier and cheaper to produce. By 1990, the desktop-publishing market had grown to $2.8 billion. The infrastructure made it happen.

But Apple was not without its problems. The market had been clamoring for more powerful Macs. Apple responded. Apple took the Mac toward the high end of the market—powerful, but expensive. The introduction of the Mac portable in 1989 was also a disappointment to the marketplace. It was too expensive and too big. It did nothing to enhance Apple's image as a product-innovative company. The product had plenty of appealing features but the competitive marketplace defined portable as "small, lightweight, and inexpensive."

IBM also had its problems in the latter half of the 1980s. Hundreds of clones had entered the PC market. It is estimated that by 1989 about 450 companies were selling PC clones. Unlike IBM mainframe computer customers, PC buyers seemed willing to move to non-IBM PC products. The PC standards were being set by Microsoft and Intel. Microsoft marketed each succeeding operating system—DOS, MS-DOS, OS/2, and Windows 3—to all takers. Likewise, Intel provided increased performance with succeeding generations of microprocessors—the 286, 386, and 486.

IBM had to do something. By staying in the Microsoft camp IBM would have to share the market with hundreds of clones, many of whom were buying their way into the market by selling them products at low prices. It was hard to make money. In late 1990, after several years and software versions, Microsoft introduced Windows 3 and it was a huge success. IBM chose to develop OS/2 with Presentation Manager instead of Windows in an apparent effort to move further away from Microsoft's control. In 1991 Microsoft announced yet another version of Windows and described the various positions targeted for OS/2 and Windows in the marketplace—IBM notwithstanding. *Infoworld* editor Michael Miller, commenting on the announcement, said, "It sounds fascinating, but it's a long way from reality. If the past is any guide to the future, it seems certain that we are going to see a lot more clarifications, retractions, and repositioning statements before the next generation operating system is available." That statement applies equally well to all facets of this industry. So stay tuned. The scenario has yet to play itself out.

In the meantime, Mac's competitive position as the friendliest computer to use was being challenged. Technology never stands still. The Mac stood still—in its original form—for too long. The development of Windows was a clear effort to bring the user interface to the PC industry. The "user-friendly gap" was closing.

In addition, the markets for personal computers have become more systems-like. That is, personal computers are now an integrated part of a business's information network. The ability to attach to a network with other types of computers and upload or download information to a central data base is essential in small and big businesses. The personal computer has become a networked computer. Field organizations are now tied into their company's central data base in order to access pricing and availability information. Distribution centers are networked across the United States and even across the world. Suppliers and customers are networked together. Even the education market is networked. Most companies want computers that can be integrated into their existing environments or easily adapted with software or peripherals for use in networks. While the Mac has one of the easiest-to-implement networking capabilities, its ability to work in multivendor computer environments still is not acknowledged by the infrastructure.

This current competitive environment is a lot different from the environment the Mac entered when it was introduced in 1984. The technology, the competition, the customer, the expectations of the market—all have changed dramatically. Mac now competes in this new environment. In order to keep up with changes in the marketplace Apple developed alternatives to its popular Macintosh. In late 1990 Apple introduced the "Classic," a low-cost Mac. The market infrastructure had been demanding such a product, and when Apple responded it was indeed rewarded. The Classic has been the largest-selling product Apple has ever introduced.

But the issues Apple had to face in order to adapt to the marketplace involved people. The early Mac people, for the most part, were not systems people. Many had never worked in the computer industry. They were evangelists. They talked more to the media and the marketing channel than to customers. They were more concerned with sending messages to the marketplace than with visiting the marketplace. They did not adapt rapidly enough to the changing marketplace and they didn't respond fast enough to the advice of the infrastructure.

While the outside world criticized John Sculley for frequent management changes and reorganizations, he was, in fact, restructuring Apple to compete in an entirely new environment. He knew that Apple had to change. In an article in the *Sloan Management Review* titled "Organizational Learning—The Key to Management Innovation," chairman and president of Analog Devices Ray Stata addresses this issue. He says, "...the rate at which individuals and organizations learn may become the only sustainable competitive advantage, especially in knowledge-intensive industries." Expectations for Apple often exceed what is possible. But Apple has continued to be a learning organization.

One change that symbolizes many of the others was naming Michael Spindler president of Apple Computer. Michael is a German citizen who spent most of his career working for Digital Equipment and Intel's international branches. He knows the computer industry and the underlying technology. He worked for Apple for more than a decade, so he knows Apple history and culture. Because of his international experience—in places from Europe to the Far East—he sees the future of Apple in a global competitive context. Apple is now a mixture of the new and the old.

The enthusiasm for innovation and industry leadership hasn't changed at Apple. But a large dose of reality has set in. The technology, markets, infrastructure, and competition all have changed and will change even more in the next decade. Apple products make up only 10 percent of the personal computer market, and Apple's challenge is to double or triple that in the next five to ten years. To succeed, Apple must be innovative, not only with the products it develops, but also in the way it does business. Apple is a more mature, competitive company today than it used to be. Of all the lessons learned from looking at the history of the Mac, the most important is that a company's vision is a lot easier to articulate than to implement. Vision doesn't create customers.

Apple is indeed a fascinating company; probably one of the most interesting, innovative, and constantly changing companies to have come onto the American scene in the past fifty years. It is a company that continues to recreate itself.

MACADDENDUM

July 3, 1991, Apple and IBM shocked the computer world by announcing that they would collaborate on technology for future computer systems. Such an alliance between two companies that had always been viewed as being at opposite ends of the cultural and technological spectrum revealed a world turned upside down. Indeed it had. The announcement said Apple and IBM would work together "to create powerful new open systems software platforms for the 1990s." The alliance was broad-ranged and involved Apple using future generations of an IBM-developed microprocessor chip. Joint ventures would be formed to develop new technologies and software products, and both companies would work together to integrate the Macintosh into the IBM client-server enterprise system environment. In layman's terms, that means the Mac will be able to work side-by-side with IBM machines in the Fortune 500 corporations. The most ambitious portions of the arrangement weren't intended to be a simple technology swap, but aim at creating a whole new computing environment. The two companies also agreed to work together to create software and multimedia technology to be licensed by other manufacturers. These products are intended to work on computers other than those produced by IBM and Apple. At that point, the agreement and the strategy extended its reach well beyond the two companies.

When the Mac was introduced in 1984, an Apple-IBM scenario was unimaginable. We must keep reminding ourselves that the personal computer industry is about 15 years old and has only reached its adolescence. It is characteristic of this age for surprises and dramatic changes to occur.

Various market conditions, changing technology, and opportunity led these two companies to link their fortunes. The movement to open systems and open standards is creating an environment in which only those who control the computer architectures make money. This reality is best illustrated by a *New York Times* headline appearing two weeks after the Apple-IBM announcement. It read "Loss for Apple; Record Profit at Intel." Intel has been the number one supplier of microprocessor chips—the brains of the computer—to PC manufacturers for the last decade. Intel and Microsoft (the other profit leader in the computer industry) own, or control, the architecture for the majority of personal computers sold worldwide. Intel provides microprocessor chips and Microsoft provides operating systems software. It is in these companies' best interest to integrate more and more of the computers' tasks or functions into the chips and the information systems software, thereby adding value to their products while offering more functionality at less cost to their customers. Equipped with these products, computer manufacturers can build very sophisticated computer systems at a very low cost. Manufacturers essentially become components packagers rather than technology developers. These companies are known as the clones. In 1991, the clones owned over 60 percent of the market share in personal computers. Hundreds of companies around the world buy chips from Intel and software from Microsoft, adding little or no differentiation to the products. As a result, they beat each other's brains out on price. This intense competition has caused a continued price decline in personal computers. Today, over a hundred different suppliers offer very powerful computers for under $1,000.

As a result, profits in the $70 billion computer industry have tumbled, falling 52 percent from 1989 to 1990 and another 25 percent from 1990 to 1991. One of the major reasons for the Apple-IBM alliance is to bring some order and profit back to the computer business. Both companies believe they can create a "camp," or a group of supporting companies, whose members continually add value through technology innovation.

Apple and the Macintosh have enjoyed enormous success since 1984. The basic product concept hasn't changed much. The current version has more functions, more and better software, and faster performance, but the revolutionary "ease of use" feature still keeps the Mac the top choice among computer buyers. Steve Jobs's dream of building a computer that anyone can learn to use in 20 minutes was a hurdle few other personal computers were able to achieve. But the competing technology from Microsoft is catching up, and Apple can no longer command as large a premium for its unique technology. Oddly enough, Apple and IBM find themselves with the same problem: Their positions and technologies are being eroded by followers. Their challenge, then, is to create an industry environment in which R&D dollars can be leveraged into profits, providing more functionality at less cost to users while safeguarding their technologies and their customers from the clones.

Both companies, however, have invested more and more in R&D and market development only to see the lowest bidder reap the rewards of their investment. These low bidders are companies offering little or no technical contribution. They operate in the draft of technology developers. Many argue that the consumer is the beneficiary of all these low-priced computers. That is true as long as market conditions don't drive the *contributors* out of business. Competition isn't going to disappear. In fact, there are now at least half a dozen computer camps forming. Apple and IBM are betting they can add more differentiation and value to their products and create a more successful camp than the others. Both companies feel they must have a major stake in the chip and operating systems software technology if they are to add value to their future products. It is not a risk-free plan. U.S. companies have been unsuccessful in building long-lasting alliances. IBM and Apple have taken a play from the Japanese book by sharing the cost and risk of technology development while they continue to compete head-to-head in the marketplace. I don't expect that Apple will cease to be Apple or IBM to be IBM. The big Japanese companies are distinctly different in corporate cultures and competitive postures, yet they have long shared technology and market development costs while competing fiercely in the marketplace.

The final chapter is far from certain. The 1991 announcement was made upon the signing of a letter of intent. There will be many trials, tribulations, and challenges before the complete story can be written. The 1984 Apple TV commercial that depicted an Orwellian Big Brother scene being shattered with the advent of a new computer—the Macintosh—came true. From a world in which everyone wanted to be and to do the same things as everyone else, we are entering a world in which everyone wants to be unique. The challenge for Apple and its Mac has not changed. Apple's future depends upon how well it can keep its innovation from drowning in the sea of clones.

Oddly enough, it is Apple and IBM together who are trying to bring differentiation back to the computer industry.

Twenty-First Century Marketing

The history of the Mac illustrates the evolution of marketing concepts. The Mac began with a handful of designers developing a revolutionary product in an isolated back room. The Mac was a spectacular product, and its introduction attracted a lot of attention. Seven years later, however, the battle for positioning goes on. Apple is still evolving and adapting the Mac to the marketplace. Today, Apple's customers are making as much impact on the Mac's evolution as the technical people. Products are no longer created in a vacuum.

When the Mac was introduced eight years ago, the personal computer revolution was in its infancy. Personal computers were individual productivity tools, viewed as sophisticated replacements for typewriters. True, the personal computer could do a lot more than the typewriter and it made work a lot easier, but it was not considered a strategic information tool. All that has changed. Personal computers have not only become more powerful, they have become an efficient way to communicate information. Corporations, big and small, can't operate without their computer networks.

As we move toward the twenty-first century, computer technology is changing even more rapidly. The power in a desktop computer circa 1990 will be available in a device that fits in the palm of your hand by the year 2000. It has long been held that information can be used as a competitive weapon. Unfortunately, the way information was gathered and stored kept it inaccessible to most of the organization. Merging information and communication in the personal computer has changed all that. As the personal computer evolves in the next decade, it will be possible to carry on a real-time dialogue between customer and supplier. Information will be communicated back and forth, creating a new marketing environment.

What does this new marketing environment mean for the future of marketing and salespeople? Companies often welcome new sales tools because increased sales means increased profits. While sales and marketing functions constantly adapt to modernization, old techniques are not always abandoned. Selling has thrived because it has diversified along with the markets it serves. As technology has created a vast array of products and options, global markets, and niche markets, it has also fragmented and expanded the salesperson's role.

In the mid-1980s, salespeople began to adopt personal computing software as another tool in their problem-solving package. A computer program might help a salesperson track and analyze customers' buying patterns, allowing him to get a jump on the most appropriate products and services to offer.

We have seen how Proctor & Gamble to better target its marketing strategies entered into a joint venture with Metaphor Computer, a software company. Metaphor software performs an invaluable service for the consumer products giant by gathering information from bar codes and checkout scanners that P & G market researchers can use quickly to adjust price, coupon, point-of-purchase display, and sales strategies. Before this system was available, it took weeks to gather, analyze, and respond to such data with adjustments in marketing strategies. With this timely, detailed information, P & G can divide markets into infinitesimal segments, even knowing exactly how much prospective customers will be willing to pay in a specific store. P & G also has a no-clip coupon system that gives discounts at the cash register and prints out P & G coupons for the next shopping trip. This type of precisely aimed micromarketing is still in its infancy, however, not for lack of data or technology but for lack of timely, imaginative ways to use it.

The next step after micromarketing is custom design, that is, one-of-a-kind products or systems. In the computer industry, the need for custom-designed systems is so great that it has created a sub-industry of systems integrators, middlemen of sorts who—by putting together computer hardware and software components from different manufacturers—tailor-make systems to fill clients' individual business requirements.

Custom-tailored services are available to consumers as well. With interactive computing technology, for example, a product called Bookseller's Assistant eliminates the need for a living, breathing salesperson by helping customers find suitable reading material. A browser enters into the system titles of their favorite movies, books, and magazines. Based on that information, the computer prints out a customized list of books that would most likely be enjoyed.

Technology gives consumers enormous choice in the design of what they buy. In the world of fashion, we can have our images copied and projected on a monitor. Our images can "try on" unlimited cuts and colors of business suits and we can judge how we look without ever stepping into a dressing room. We can do the same with makeup, experimenting with a rainbow of looks on our computer image, all the while avoiding the cotton puffs and cold cream it takes to experiment the old-fashioned way. Using the measurements of our existing kitchens, we can add counters, cabinets, and floors to the three-dimensional computer image, making design decisions without having to knock down a wall or move a stick of furniture. We can choose paints with hues that are computer-matched to the flowers on our couches or design the color of our carpets to match a favorite color plucked from memory.

A revamped, formerly low-tech trucking industry offers customers computerized access to shipments. Skyway Freight Systems of Santa Cruz tracks shipments via computer, then lets customers in on the company's systems to monitor their own shipments without middleman delay. It also offers a computerized just-in-time delivery system similar to the shipping system that has allowed Japanese manufacturers to keep tight control of shipping and inventory costs. JIT shipping allows companies to hold down these costs by insuring that a shipment gets to the plant just as the production process demands it, but not faster than workers can handle it.

The complexity, cost, and risk of technology products makes them fundamentally different from consumer products, and marketing practices are usually different as well. There are points of convergence, however, the most important of which is the need for experiential marketing, encompassing the variety of new ways in which a customer can experience a product or service.

In consumer products, there are numerous examples of experiential marketing: free samples of Mrs. Fields cookies, perfume-laced inserts in magazines, free soap in the mail, and test-drive offers that auto dealers use to draw customers.

In markets that are highly competitive and quickly changing, salespeople have to come up with imaginative experiential marketing techniques. For example, potential customers are offered trial software that lets them preview goods before buying. Word-of-mouth references are another type of experiential marketing. A potential customer is likely to give special weight to the experiences of another buyer and inferentially apply the latter's experiences to his own situation. Yet another new and powerful technique, which takes advantage of the flexibility of new technology, is to involve the customer in the design of the product, be it custom-designed chips, software, or networking systems. A variation on this was Genentech's decision to involve prominent cardiologists in testing TPA, an anti-clotting drug, which paved the way for the drug's early success following FDA approval.

The significance of experiential marketing cannot be overstated. Lack of it figures prominently in why large, highly bureaucratized institutional firms run into trouble: They have gotten too far away from their customers' needs and desires. Throwing millions of dollars into advertising, for instance, is a waste. Advertising does not sell a product; references and reputations do. At best, ads can reinforce positive beliefs about a product and company based on what the customer has already heard or directly experienced. Advertising cannot overcome customer resistance based on more immediate experiences.

We can already glimpse the future for high technology selling. It includes voice input sales where customers use television, telephone, and computerized "salespeople" to do their in-home shopping. A customer can view a product on TV, call in his order, and dictate to an interactive computer the features he wants, which helps him narrow his choice to the ideal purchase. If the purchase is, say, a washing machine, and it breaks down, computerized customer service would allow the owner to telephone a service center, enter the diagnostic code she sees flashing on the machine, then follow computerized instructions for making simple repairs.

Marketing and salespeople of the new century will not recognize their forebears of the twentieth century. The marketeer and salesperson of the future will have to be as adept at information technology as the design engineer of today. In the twenty-first century, I believe the marketing and the sales professions will be held in the same esteem as the science and engineering professions of today.

| # Test Your Marketing IQ

Questions

1. How many customer visits has your CEO made in the past year?

   Fewer than 5 ☐   5–10 ☐   10–15 ☐   15–20 ☐   More than 20 ☐

2. How often do marketing, manufacturing, and engineering managements meet to discuss customer feedback?

   Once per quarter ☐   Twice per year ☐   Rarely ☐   Never ☐

3. How often do members of marketing, manufacturing, and engineering departments meet to discuss ways to adapt current products and services to current market needs?

   Once per quarter ☐   Twice per year ☐   Rarely ☐   Never ☐

4. Rate your company's "marketing value" in determining product direction.

   Weak ☐   So-so ☐   Pretty good ☐   Excellent ☐   Don't know ☐

5. The ideal marketing plan involves researching the market, defining the need, then telling engineering what to build.

   True ☐   False ☐

6. Do you have specific user-group or customer-group meetings for the purpose of gaining customer feedback?

   Yes ☐   No ☐

7. Who makes new-product decisions?

   The Guru ☐   Engineering ☐   Engineering and Marketing ☐

8. Do you involve customers early in the product-design cycle?

   Yes ☐   No ☐   Sometimes ☐

9. How would you rate your company on building relationships (with R&D people, OEM people, suppliers, etc.)?

   Poor ☐   So-so ☐   Not too bad ☐   Good ☐   Excellent ☐

10. Would you characterize your company as having "NIH" (Not Invented Here) mentality?

    Absolutely ☐   Somewhat ☐   Not at all ☐

11. How many sentences would it take you to explain your company's strategic marketing direction?

    Fewer than 3 ☐   5–10 ☐   More than 10 ☐

12. How much time do your management people spend discussing market strategies?

    Quite a bit ☐   A fair amount ☐   Some ☐   Very little ☐

13. How much time of your company's executive staff meeting is devoted to discussion of marketing trends, influences, or new directions?

    5–10% ☐   10–15% ☐   15–20% ☐   More than 20% ☐

14. How does your company define marketing? (Pick one.)

    A. As sales and promotion
    B. As identifying future markets and customers
    C. As interaction between engineers and customers
    D. As guiding the adaptation process of products to market
    E. As relationship building

15. Your customers' perceptions are reality.

    Agree ☐   Disagree ☐

16. Advertising can create markets for complex products.

    Agree ☐   Disagree ☐

17. How would you rank your company on the service spectrum?

    Low ☐   Medium ☐   High ☐   It's all that counts ☐

18. A company's position can be best defined by a good, crisp slogan.

    Agree ☐   Disagree ☐

19. Incremental product improvement and innovation is the prescription for success.

    Agree ☐   Disagree ☐

20. A company's financial position has little or nothing to do with the company's market position.

    Agree ☐   Disagree ☐

21. Which of the following is the most credible (and therefore most effective) communication method for establishing a new product? (Pick one.)

    A. Advertising
    B. Direct Mail
    C. Sales Calls
    D. Seminars
    E. Word of Mouth

22. Being first to market assures a leadership position.

    Agree ☐   Disagree ☐

23. Niche markets are more difficult to support than general purpose, broad markets.

    Agree ☐   Disagree ☐

24. Leadership companies don't seek niche markets because return on investment and profitability are much lower there than can be achieved in the broad marketplace.

    Agree ☐   Disagree ☐

25. A good advertising or marketing communications objective is to increase sales X percent over the previous year's sales.

    Agree ☐   Disagree ☐

Answers and Score

1. Decision making in fast changing markets requires experienced judgement. CEO's in technology businesses must keep in touch with their customers' trends, needs, and problems.

   Fewer than 5 customer visits (2 pts);   5–10 (5 pts);   10–15 (6 pts); 15–20 (8 pts);   More than 20 (20 pts)

2. Every aspect of business has an influence on every other aspect. Flexible organizations understand that the roles and capabilities of all parts of the organization are interrelated.

   Once per quarter (5 pts);   Twice per year (2 pts);   Rarely (0 pts); Never (0 pts)

3. All successful technology products are adapted to specific markets or are altered to meet competitive challenges.

   Once per quarter (5 pts);   Twice per year (2 pts);   Rarely (0 pts); Never (0 pts)

4. More often than not, marketing is closer to the customer than is engineering or other management. Competent marketing people take a leadership role in the direction of new-product development.

   Weak (0 pts);   So-so (0 pts);   Pretty good (2 pts);   Excellent (5 pts); Don't know (0 pts)

5. The ideal marketing plan comes about by having one eye on advances in technology and one eye on the market. It takes experienced insight and judgement to assess the capabilities of the technology and the opportunities of the marketplace.

True (0 pts);   False (5 pts)

6. With the market environment constantly changing, on-going relationships with key user groups is one sure way to keep your products on target.

Yes (4 pts);   No (0 pts)

7. The size of the company often dictates who makes product decisions. As more than one product emerges, it is necessary to have in place a process of checks and balances in order to assure that the market has its say.

The Guru (1 pt);   Engineering (2 pts);
Engineering and Marketing (4 pts)

8. Customer involvement in the early stages of development not only assures early sales of the product, but such practices help to eliminate bugs, enhance the product, and establish a base for the development of word-of-mouth communication.

Yes (5 pts);   No (0 pts);   Sometimes (0 pts)

9. No one can go it alone in the complex businesses of technology. Even IBM, with all its resources, has increased its programs for building strategic relationships.

Poor (0 pts);   So-so (0 pts);   Not too bad (2 pts);   Good (3 pts);
Excellent (4 pts)

10. Complex products are composed of interrelated technologies. Keeping pace in all areas of business and technology is an almost impossible task.

Absolutely (4 pts);   Somewhat (2 pts);   Not at all (0 pts)

11. Writing your strategic marketing direction in one sentence or ten is not the issue. The real question is whether an immediate concept came to mind when you read the question. Do you have a strategy?

    Fewer than 3 (4 pts);   5–10 (3 pts);   More than 10 (1 pt)

12. Management is often driven by sales and factory reports, which consume most of the staff meeting time. In reality, the customer makes the sales and the factory tick, not vice versa.

    Quite a bit (5 pts);   A fair amount (4 pts);   Some (2 pts);
    Very little (0 pts)

13. Most managements spend lots of time discussing reports and analyses, rather than evaluating the environment.

    5–10% (2 pts);   10–15% (3 pts);   15–20% (4 pts);
    More than 20% (5 pts)

14. There are many definitions for marketing, but no doubt the most valuable way of defining marketing is as "relationship building." If marketing people build solid market relationships, then their knowledge of the customer and communication with the customer is a constant value. The same holds true with dealer, distributor, media, and engineering relationships.

    A. As sales and promotion (0 pts);   B. As identifying future markets and customers (0 pts);   C. As interaction between the engineering department and the customer (2 pts);   D. As guiding the adaptation process of products to market (4 pts);   E. As relationship building (5 pts)

15. If reality and perception are not the same thing, then it's because you have not done a good job of producing tangible evidence for your position and communicating the value of your product or service to the market.

    Agree (2 pts);   Disagree (0 pts)

16. Complex products and most services are best sold by word of mouth and a reference strategy. Advertising can build awareness, but it can rarely communicate the same level of confidence as delivered by a reference system.

    Agree (0 pts);   Disagree (2 pts)

17. Service is an integral part of the product's definition. Further, service has become a way of differentiating products and building strong relationships with customers.

    Low (0 pts);   Medium (0 pts);   High (3 pts);
    It's all that counts (5 pts)

18. A great deal of money is wasted in both producing and promoting slogans. Studies show that few slogans are even remembered or correctly associated with the right company. A position is something your customers see in you, not something you want to be.

    Agree (0 pts);   Disagree (2 pts)

19. If you bring out products that the market wants, you are in a position to continually innovate. Product improvement and innovation is a leading indicator of success.

    Agree (2 pts);   Disagree (0 pts)

20. A poor financial position sends signals to the market. Corporate visibility, even for nonpublic companies, is very high. Poor financial performance raises concerns about the future of the company, the competitiveness of its products, and the quality of its management.

    Agree (0 pts);   Disagree (3 pts)

21. Professional services have used word of mouth because trust and confidence are of primary importance. When the product or service is central to the strategy and performance of the customer, confidence and trust are the two most important assets one can convey, especially in highly competitive markets involving many players and lots of activity. Those two qualities can be communicated only through an experience-based reference system.

    A. Advertising (0 pts);   B. Direct Mail (0 pts);   C. Sales Calls (2 pts);
    D. Seminars (2 pts);   E. Word of Mouth (5 pts)

22. Pioneers still get arrows in their backs. Being first to market is an advantage if technology leadership can be maintained for some time, if the company has the resources to develop and support the market, and if

it continues to innovate. But being first to market is no guarantee of success.

Agree (0 pts);   Disagree (3 pts)

23. While niche markets are often thought of as safe havens, they are, in fact, more demanding. Customer needs must be met and supported more precisely and the reference system must be more tightly coupled, so your reputation is of prime importance.

Agree (3 pts);   Disagree (0 pts)

24. Studies show that companies achieving leadership in niche markets have a higher return on investment and profitability than those going for market share in established, broad markets.

Agree (0 pts);   Disagree (3 pts)

25. The purpose of advertising is to achieve awareness. Unfortunately, awareness is not synonymous with behavior change. The expectation that advertising or public relations can manipulate the potential buyer is like putting all the emphasis on the component rather than on the system.

Agree (0 pts);   Disagree (3 pts)

Scoring:

> 85–100: Genius, Marketing-Driven Company
> 75–85: Smart, Marketing Champion
> 65–75: No Cigar
Below 65: Lots of Room to Grow

Reading List

The following list contains some of my favorite books and papers. It is not comprehensive nor does it contain only books on the subject of marketing. I have found, in the course of my career, that an awareness and study of people, history, political issues, and social and technology trends all lead to a better understanding of the dynamics of the marketplace.

James C. Abegglen and George Stalk, Jr., *Kaisha: The Japanese Corporation,* Basic Books, 1985.

Daniel J. Boorstin, *The Discoverers: A History of Man's Search to Know His World and Himself,* Vintage Books, Random House, 1985.

Frederick P. Brooks, Jr., *The Mythical Man-Month: Essays on Software Engineering,* Addison-Wesley, 1982.

Robert A. Burgelman and Leonard R. Sayles, *Inside Corporate Innovation: Strategy, Structure and Managerial Skills,* The Free Press, 1986.

Donald K. Clifford, Jr., and Richard E. Cavanagh, *The Winning Performance: How America's High-Growth Midsize Companies Succeed,* Bantam Books, 1985.

E. Raymond Corey, Frank V. Cespedes, and V. Kasturi Rangan, *Going to Market: Distribution Systems for Industrial Products,* Harvard Business School Press, 1989.

William H. Davidow, **"Marketing High Technology: An Insider's View,"** *Free Press,* June 1986.

Peter F. Drucker, **"The Effective Decision,"** *Harvard Business Review,* January–February 1967.

Peter F. Drucker, *The New Realities,* Harper and Row, 1989.

Freeman Dyson, *Infinite in All Directions,* Harper & Row, 1985.

Stuart Ewen, *All Consuming Images,* Basic Books, 1988.

Paul Freiberger, *Fire in the Valley,* Osborn-McGraw, May 1984 (out of print).

O. B. Hardison, Jr., *Disappearing Through the Skylight: Culture and Technology in the Twentieth Century,* Viking Penguin, 1989.

John K. Johansson and Ikujiro Nonaka, "**Market Research the Japanese Way,**" *Harvard Business Review,* May–June 1987.

Michael Kammen, *People of Paradox: An Inquiry Concerning the Origins of American Civilization,* Oxford University Press, 1980 (paperback).

Rosabeth Moss Kanter, *When Giants Learn to Dance,* Touchstone: Simon & Schuster, 1989.

Tracy Kidder, *The Soul of a New Machine,* Avon Books, 1981.

Theodore Levitt, "**Branding on Trial,**" *Harvard Business Review,* March–April 1966.

Theodore Levitt, *The Marketing Imagination,* The Free Press, 1986.

Theodore Levitt, "**Marketing Intangible Products and Product Intangibles,**" *Harvard Business Review,* May–June 1981.

Theodore Levitt, "**Marketing Myopia,**" *Harvard Business Review,* September–October 1975.

Theodore Levitt, *Thinking about Management,* The Free Press, 1991.

Edwin Mansfield, "**Industrial Innovation in Japan and the United States,**" *Science,* Vol. 241, September 30, 1988.

Regis McKenna, "**Marketing in the Age of Diversity,**" *Harvard Business Review,* September–October 1988.

Marshall McLuhan, *Understanding Media: The Extensions of Man,* Signet Books, 1964.

Rowland T. Moriarty and Thomas J. Kosnik, "**High-Tech Marketing: Concepts, Continuity, and Change,**" *Sloan Management Review,* Summer 1989.

Rowland T. Moriarty and Gordon S. Swartz, "**Automation to Boost Sales and Marketing,**" *Harvard Business Review,* January–February 1989.

Heinz R. Pagels, *Dreams of Reason: The Computer and the Rise of the Sciences of Complexity,* Bantam, 1989.

Richard Tanner Pascale, *Managing on the Edge: How the Smartest Companies Use Conflict to Stay Ahead,* Simon & Schuster, 1990.

Tom Peters, *Thriving on Chaos,* Alfred A. Knopf, Inc., 1987.

Michael E. Porter, *Competitive Strategies: Techniques for Analyzing Industries and Competitors,* The Free Press, 1980.

Michael E. Porter, **"From Competitive Advantage to Corporate Strategy,"** *Harvard Business Review,* May–June 1987.

Michael E. Porter, **"How Competitive Forces Shape Strategy,"** *Harvard Business Review,* March–April 1979.

Nathan Rosenberg, *Inside the Black Box: Technology and Economics,* Cambridge University Press, 1982.

Nathan Rosenberg and L.E. Birdzell, Jr., *How the West Grew Rich: The Economic Transformation of the Industrial World,* Basic Books, 1986.

Rudy Rucker, *The Fourth Dimension: A Guided Tour of the Higher Universe,* Houghton Mifflin, 1984.

E. F. Schumacher, *Small Is Beautiful: Economics As If People Mattered,* Harper & Row, 1989 (reissued).

William L. Shanklin and John K. Ryans, Jr., **"Organizing for High-Tech Marketing,"** *Harvard Business Review,* November–December 1984.

G. Lynn Shostack, **"Breaking Free from Product Marketing,"** *Journal of Marketing,* April 1977.

George Stalk, Jr., **"Time: The Next Source of Competitive Advantage,"** *Harvard Business Review,* July–August 1988.

Ray Stata, **"Organizational Learning: The Key to Management Innovation,"** *Sloan Management Review Reprint Series,* Spring 1989.

Richard S. Tedlow, *New and Improved: The Story of Mass Marketing in America,* Basic Books, 1990.

Alvin Toffler, *Power Shift,* Bantam Books, 1990.

Barbara W. Tuchman, *The March of Folly: From Troy to Vietnam,* Ballantine Books, 1985.

Index

Abbott, Edwin, 71
Acquisition strategies, 106–7
Activision, 154
Acura, 74
Adaptation, 4–5, 9, 79; and advertising, 13; and bringing new users into the market, 77; and experience-based marketing, 14; and incremental improvements, 77–78; "sequence," 27, 114–16; and success, 75–76
Adobe, 28, 209
Advertising, 14, 21, 23, 28, 204–7, 215, 219; and bigness mentality, 181; and broken chains, 182; and credibility, 87; and customer mentality, 177; glut of, 12–13, 33; as an indirect form of dialogue, 119; and the "marketeer," 52; and the old notion of marketing, 5, 11–12; as only a small part of marketing strategy, 32; and positioning strategies, 45; and product availability, 125–26; research, 151; slogans, 45–46; television, 12, 13, 31, 204–7, 215; testing of, 18
Aldus, 209
Alliances, 56, 102–3. *See also* Joint ventures; Strategic alliances
AM International, 106
Amdahl, Gene, 110, 137
American Can, 51
American Express, 87
Analog Devices, 188, 211
Anderson, Howard, 51

Antitrust, 111–12
Apple Computer, 28, 122, 216; Apple II, 38–39, 72–74, 78, 85, 179–80, 190–91; Apple III, 85; "Classic," 211; and credibility, 196, 198, 199; and IBM, collaboration with, 212–15; and "industry watchers," 93–94; laser printer, 209; LISA, 27, 80, 191, 194–98, 201–3, 206, 209; Macintosh, 13–14, 74, 189–216; Mac portable, 210; and the workings of the environment, 38–39
Appliance, use of the word, 194
Applied Materials, 33–34, 87
Aristotle, 176
Arm & Hammer, 60–61
S. Armacost, 65
Arthur Rock, 142
ASICs, 110
ASK Computer, 114
AT&T, 32–33, 87, 116; and applied research, 107; and the product concept, 184; and small companies, deals with, 88
Atari, 173, 196, 198
ATM machines, 66
Audiences, 76–77, 84, 192–95
Audits: external, 146, 150–55, 163, 165–66, 199–200; internal audits, 144–50
Automobile industry, 51–52, 203; and automatic transmissions, 77, 78; and commodity mentality, 179; and foreign competition, 38, 74, 171, 173, 175; and the Japanese, 38, 74,

Yankee Group, 51
Yankelovich Clancy Shulman, 3

Yellow Pages, 32
Yuppies, 11